CHASING THE GHOST

Nobelist Fred Reines and the Neutrino

Cover image: Super-Kamiokande Neutrino Detector.
Courtesy of the Kamioka Observatory, ICRR (Institute for Cosmic Ray Research)
The University of Tokyo

The Super-K is a huge tank in a deep underground Japanese cavern. Surrounded by light sensors, it is 13-stories tall and holds 50,000 tons of ultra-pure water. The Super-K and other giant neutrino detectors are successors to the far smaller device first created by Fred Reines to discover the ghostlike particle. Takaaki Kajita, like Reines a Nobel Laureate, received the Prize for his neutrino findings on the Super-K, and he pays homage to Reines' pioneering accomplishments: "This book," he says about *Chasing the Ghost*, "is a most welcome account about Frederick Reines and his great contributions to neutrino physics and astrophysics."

CHASING THE GHOST

Nobelist Fred Reines and the Neutrino

Leonard A Cole

Rutgers University, USA

World Scientific

NEW JERSEY • LONDON • SINGAPORE • BEIJING • SHANGHAI • HONG KONG • TAIPEI • CHENNAI • TOKYO

Published by

World Scientific Publishing Co. Pte. Ltd.
5 Toh Tuck Link, Singapore 596224
USA office: 27 Warren Street, Suite 401-402, Hackensack, NJ 07601
UK office: 57 Shelton Street, Covent Garden, London WC2H 9HE

Library of Congress Control Number: 2021934123

British Library Cataloguing-in-Publication Data
A catalogue record for this book is available from the British Library.

CHASING THE GHOST
Nobelist Fred Reines and the Neutrino

ISBN 978-981-123-105-6 (hardcover)
ISBN 978-981-123-148-3 (paperback)
ISBN 978-981-123-106-3 (ebook for institutions)
ISBN 978-981-123-107-0 (ebook for individuals)

For any available supplementary material, please visit
https://www.worldscientific.com/worldscibooks/10.1142/12125#t=suppl

Typeset by Stallion Press
Email: enquiries@stallionpress.com

Foreword

Douglas R. Hofstadter

Distinguished College of Arts and Sciences Professor
Indiana University, Bloomington

I'm sad that I met Frederick Reines on only two occasions, but so grateful for the powerful and indelible impression that he left on me. Every time I remember him or hear about him, I feel a sense of admiration and an irrepressible urge to smile. He was, as they say in Yiddish, a real *mensch* — a deep, generous, and always reliable soul.

Fred Reines was one of the twentieth century's true pioneers in physics. He had great visions and great ambitions, and yet he was astonishingly modest. His most significant achievement was probably his "sighting" of the extraordinarily elusive neutrino in the mid-1950s, carried out with his friend Clyde Cowan. This particle had been hypothesized in 1930 by the Austrian theorist Wolfgang Pauli, who was so ashamed of his idea that he didn't dare write an article about it, or even present it in a talk. Instead, he typed a humorous letter to be read aloud by a friend at a conference, in which he semi-seriously suggested a new particle that, if it existed, could save the sacrosanct laws of conservation of energy and momentum, both of which seemed to be violated by radioactive decays — but Pauli also apologized for his brashness in making such a wild hypothesis (a "desperate remedy", as he called it). Four years later, two famous theorists — Hans Bethe and Rudolf Peierls — wrote an article in which they stated that the neutrino, should it by chance exist (which was far from clear), would "obviously" never be detected by human beings.

This kind of throwing-down of the gauntlet was irresistible to Fred Reines, and in the late 1940s and early 1950s, he began formulating possible strategies to "see" the "unseeable" neutrino (if indeed it existed!). His first brainstorm involved exploding a nuclear bomb, with an underground detector only 50 meters (!) from the explosion point. (Some wag suggested that a graduate student should be placed right next to the detector, in order to babysit it. The suggestion was politely vetoed.) Reines and his colleague Cowan worked on this idea for a couple of years, but eventually they were talked out of it, deciding that it would make more sense to use a nuclear reactor as a potential neutrino source, even though it was far less intense a source. At the time, the reality of neutrinos was still far from certain, so Reines and Cowan light-heartedly dubbed their experiment "Project Poltergeist" (a poltergeist being an invisible spirit that rumbles around making strange noises), and they gave their detector the name "Herr Auge" ("Mr. Eye", in German).

The story of how Reines and Cowan finally did indeed "see" the neutrino (that is, "see" it through a highly recognizable kind of "signature" written in the language of photons) is admirably told in this loving biography by Fred's younger cousin Leonard Cole, who as a boy was inspired by his affectionate cousin to explore science and to savor its beauty and depth. Cole went on to become first a dentist, then a political scientist, and then a world expert on bioterrorism. But he never forgot the influence of his beloved older cousin, and later in life he decided that Fred Reines deserved a biography. This excellent decision was followed up by years of very intense work and travel. Cole visited many laboratories in the United States and Europe, and talked with dozens of colleagues, students, and family members whose vivid recollections of Fred Reines pepper this book in a very lively fashion.

What is impressive to me is how deeply Leonard Cole has internalized not only the physics itself but also the history of twentieth-century physics, both of which he relates in a crystal-clear manner to lay readers. He takes his cue from Fred Reines himself, whose *modus operandi*, as a communicator in his humbly-named course for non-physics majors "Rainbows and Things", was to use the simplest possible means of expression, bypassing mathematics and in its place using vivid images, colorful analogies, clear demonstrations with ordinary objects (candles, paper clips, pieces of string), and even singing and whistling, as well as reciting poems.

Speaking of singing and whistling, Fred Reines was a deeply musical soul, and it was with enormous gusto and panache that he performed songs and arias from such diverse sources as Gershwin, Bach, Handel, and Gilbert and Sullivan. He also loved acting, and played the role of "Drummond" (alias Clarence Darrow) in the famous drama about the Scopes "monkey trial", *Inherit the Wind*. But Reines' greatest love was the pursuit of the neutrino, which, after a while, became the pursuit of different *types* of neutrinos, and of neutrinos emanating from different *sources*, such as the sun, a supernova, or the Big Bang.

In 1962 it was discovered that there were at least two different types of neutrino, not just one, as had been originally believed, and Reines was fascinated. However, in 1988, when this later discovery was rewarded by a Nobel Prize while the discovery of the *first* type of neutrino had gone totally unrecognized, Reines was shocked and hurt. He was never one to clamor for glory, but he felt that something was deeply wrong when the *second* neutrino had been recognized by a Nobel Prize, but not the *first* one. An amazing anecdote is told in Chapter 15 of this book about a lecture at the great laboratory of CERN in Geneva in which this crazy irony was brilliantly pinpointed by the speaker. Fred was sitting in the audience right next to Leon Lederman, one of the Nobel winners for the *second* neutrino, and the speaker said.... But I will not give it away. It is just too delicious for words.

Luckily, the Nobel Committee had the wisdom, at long last, to recognize the greatness of the classic Reines–Cowan experiment, and they awarded Fred Reines a Nobel Prize for Physics in 1995, but by that time, sadly, his co-discoverer Clyde Cowan had been dead for over twenty years. Such are the unfortunate twists and turns of fate. By now, many Nobel Prizes have in fact been given for discoveries related to various facets of the neutrino (or rather, the *neutrinos*, since now there are believed to be three types, plus three *antineutrinos*, as well), and the more that has been learned about these "poltergeists", the more bizarre they seem. The lore and lure of neutrinos constitutes such an enticing and rich saga that anyone interested in ideas at all should take some time to read about it.

I am so impressed and gratified that Fred Reines' younger and ever-loyal cousin Leonard had the inspiration to take up his pen and to write this wonderful book telling all about these mysterious particles, and recounting the life of one of the great scientists of our era and one of the finest human beings I have ever had the honor to meet.

September 2020
Bloomington, Indiana

Preface

Frederick Reines, https://www.nobelprize.org/prizes/physics/1995/reines/biographical/.

The surname "Reines" is pronounced in different ways, but the version I am used to is "RYE-ness." Thus, one morning in October 1995, I was not initially stirred when a TV newscaster announced a winner of that year's Nobel Prize in physics as "Frederick Rains." Seconds later I felt a surge of excitement. When the announcer added that the prize was for the discovery of the neutrino, I realized that my cousin Fred was that person.

The award was late in coming. Fred was 77. His 1956 discovery, with his colleague Clyde Cowan, was part of our family's lore. In the decades following its detection, the significance of the subatomic particle in the composition of the universe became increasingly appreciated.

Fred was 16 years my senior. When I was a youngster, we would connect at family gatherings, often at our grandparents' house in Paterson, New Jersey. I was always happy to see him. I regarded him as a mentor, and I believe he reciprocated this unspoken understanding. He invariably asked about my progress at school and my extracurricular interests. When I was about 8, he was especially pleased when I told him I was excited by a recent visit to the Hayden Planetarium in New York City.

To me, Fred seemed a repository of energy. At one point he offered to take me to New York's Museum of Science and Industry, then in Rockefeller Center. I gladly agreed. He enthusiastically tried to explain the varied exhibits about gravity, refraction, light, and energy. In later years my interest in science remained keen, though I was drawn especially to the biological and health sciences. Although not a scientist, I have since taught and written extensively on issues of science and public policy. My training in dental medicine and my studies for a PhD in political science provided a general understanding of the basic sciences.

Some years after Fred's death in 1998, I began to reflect on his professional accomplishments, his many talents beyond science, and on his unusually engaging persona. I spoke with other relatives who shared with me their largely admiring thoughts about him. After deciding to write this book I searched through articles by and about him. I visited locations at which he lived and worked. They included Los Alamos, New Mexico, where his professional life began with the Manhattan Project; and the University of California, Irvine where he spent his last 30 years. I also journeyed to CERN (the Geneva-based European Organization for Nuclear Research), to Brookhaven National Laboratory near New York City, and to Fermilab in Illinois, whose website describes it as "the neutrino capital of the world."[1] At each place I met with physicists who knew or knew about him.

Most rewarding were my conversations with dozens of scientists and others who worked with Fred at various stages of his career. All these experiences helped shape the portrait of a fascinating man and the fascinating particle that he discovered.

This book is a mixture of memoir and biography and it tells a human story. Although informed by recognition of the importance of neutrinos, it is primarily about Fred Reines, and not technical details of neutrino physics. The science presented here is supportive to the story about Fred and aims to appeal to a general audience.

Neutrinos are abundant throughout the universe. The minuscule particles zoom in different directions, many at nearly the speed of light. They carry no electric charge and have almost no mass, and among their strange properties is their ability to travel through solid material as if it did not exist.

[1] Fermilab, Neutrinos, http://www.fnal.gov/pub/science/particle-physics/experiments/neutrinos.html.

Neutrinos are now regarded as essential building blocks "for the blueprint of nature."[2] Many contemporary books and articles by physicist-writers refer to the neutrino's singular appeal. Frank Close has written: "Of all the things that make the universe, the commonest and weirdest are neutrinos." Similarly, from Ray Jayawardhana: "Neutrinos [are] by far the most elusive and the weirdest of all known denizens of the subatomic world." Neil deGrasse Tyson, an acclaimed popularizer of astrophysics, has called the neutrino "the most common and the most mysterious particle in the cosmos."[3]

Others have been absolutely smitten by the subject. Physicist Lawrence Krauss titled an article, "Why I Love Neutrinos," in which he wrote:

> It never ceases to amaze me that every second of every day, [trillions of] neutrinos coming from nuclear reactions inside the sun whiz through my body, almost all of which will travel right through the earth without interruption. But I am even more amazed that in spite of their ghostliness, we can detect them, probe them and unravel their mysteries.

For similar reasons, Leon Lederman, a legendary particle physicist and Nobel Prize winner, said, "Neutrinos are my favorite particles." This preference ironically came from the man whose book on the Higgs boson was titled (at his publisher's insistence), *The God Particle*.[4]

Janet Conrad, a leading neutrino physicist, concurred with an interviewer's observation that, "if the universe had no neutrinos, we wouldn't be here either. And yet they are the least understood subatomic

---

[2]Fermilab, Dan Hooper, https://www.pbs.org/newshour/science/what-is-a-neutrino-and-why-should-anyone-but-a-particle-physicist-care.

[3]Frank Close, *Neutrinos*, NY: Oxford University Press, 2010; Ray Jayawardhana, *Neutrino Hunters: The Thrilling Chase for a Ghostly Particle to Unlock the Secrets of the Universe*. NY: Scientific American/Farrar, Straus and Giroux, 2013; Neil deGrasse Tyson, *Cosmos*, https://www.youtube.com/watch?v=DOzDqLFVwsM.

[4]Lawrence M. Krauss, "Why I Love Neutrinos," *Scientific American*, June 1, 2010, https://www.scientificamerican.com/article/why-i-love-neutrinos/; Leon Lederman, *The God Particle*, NY: Delta Publishing, 1993, https://faculty.washington.edu/lynnhank/Lederman.pdf.

particle around." Conrad responded: "That's the reason I got interested in neutrinos."[5]

Fred Reines wrote a foreword to *Spaceship Neutrino*, a book by Christine Sutton. He praised her description of "the flowering of neutrino physics [which] has made it an exciting field of investigation." Sutton, a physicist and science writer, could hardly have displayed more exuberance about her topic:

> Neutrinos are for me the most fascinating aspect of a fascinating subject. They are the most amazing, paradoxical, intriguing — choose whatever adjective you like — of all elementary particles.[6]

Janet Conrad Christine Sutton
www.worldsciencefestival.com. www.physicsworld.com.

Fred thought it "an absurdity" that a purportedly educated person not know the physical basis of a rainbow or the essential ideas behind special relativity. He taught a course on physics to college non-science majors that included information about the neutrino. In that spirit my aim is to attract readers who may never have studied physics and know little or nothing about the neutrino. He believed that many fundamental concepts about science could be conveyed in everyday language. No need for complicated mathematics. Consistent with Fred's approach, I too strive for comprehensibility to the general reader.

---

[5] Dana Mackenzie, "The Neutrino Hunter," *Swarthmore College Bulletin,* January 2009, bulletin.swarthmore.edu.
[6] Christine Sutton, *Spaceship Neutrino*, NY: Cambridge University Press, 1992: xiii.

Detecting neutrinos that can pass through Earth as though it were not there, is challenging. They are identifiable by bursts of light emitted in a detector during a reaction with other particles. But those reactions are extremely rare, and neutrinos almost always pass through detectors unnoticed. Cosmic rays are also constantly bombarding the planet. Their light signals in a detector can resemble those caused by the very infrequent neutrino interactions. Efforts to block the unwanted cosmic ray background have sometimes taken scientists to formidable locations. Neutrino observatories have been built inside a mountain, deep in the Antarctic ice, at the bottom of gold, silver, and nickel mines.

In their quest to detect neutrinos, scientists may encounter unanticipated difficulties. These may arise from obstacles that seem unrelated to the intended scientific venture. My first chapter describes the effort to build a neutrino detector at the bottom of a gold mine. The chapter is less about science than a display of adverse social, political, and physical conditions encountered by Fred and other members of his research team. Once construction was completed, the experiment could begin. The scientific findings became clear in 1965 and are reported in Chapter 10, when Fred was approaching the end of a seven-year tenure at Case Institute in Cleveland.

The quest for knowledge about the wondrous universe is a profoundly human endeavor. Fred Reines' life and his unremitting scientific curiosity are emblematic of that truth.

Contents

Part I

Beginnings

Chapter 1

Challenge, Determination, and Thinking Big

In the summer of 1963, an unusual visitor entered the East Rand South African gold mine. Fred Reines, an American physicist, had come from the other side of the world with an experiment in mind. He was accompanied into the mine by an American engineer friend, Gus Hruschka, and a South African physics professor, Friedel Sellschop. Precaution for their safety was meticulous. Unlike the daily mine workers, they changed into loose-fitting coveralls to ward off dust. They wore water-repellent high-laced shoes for walking through the puddled underground passageways, and hard hats to protect against fragments bursting from the surrounding rock. Lamps affixed to their hard hats would cast beams into the dark tunnels. They also carried gas masks, in case a sensor signaled the presence of carbon monoxide or methane.

Seven years earlier, in 1956, Reines and his colleague Clyde Cowan discovered a subatomic particle, the neutrino. Although plentiful throughout the universe, neutrinos rarely interact with matter. They usually travel unimpeded and unrecognized. On average, a speeding neutrino could pass through a light-year of lead — 6 trillion miles thick — and not be slowed.[1] The minuscule particle is commonly described as elusive. More than half a century after its detection the neutrino is still referred to as a ghost particle. For example, a recent cover story in *Science* magazine about neutrino scattering was

[1] "What Is a Neutrino?" *T2K Experiment* http://t2k-experiment.org/neutrinos/what-is-a-neutrino/; https://snews.bnl.gov/popsci/neutrino.html.

headlined, "Spotting a Ghost." An article in the *Smithsonian* was titled, "Looking for Neutrinos, Nature's Ghost Particles."[2]

At the time of their discovery, Reines and Cowan were working at the Los Alamos National Laboratory in New Mexico. The neutrinos that they detected were produced by a nuclear reactor. Their discovery was hailed as a notable advance in understanding the universe. Willard Libby, chairman of the United States Atomic Energy Commission, which had funded the experiment, called the finding a "magnificent accomplishment."[3] Reines left Los Alamos in 1959 to become chairman of the physics department at Case Institute of Technology in Cleveland, Ohio, where he began to seek a location to detect neutrinos emanating from sources beyond the Earth. Confirming their existence would advance his initial "magnificent accomplishment."

Reines tried to detect atmospheric neutrinos in a salt mine 1800 feet deep, 30 miles from Cleveland. Work in the mine advanced understanding of some cosmic ray activity, but the effort to detect neutrinos was unsuccessful. The depth was not sufficient to prevent considerable cosmic radiation from reaching the underground laboratory. The cosmic interference produced signals that could mimic those from neutrinos. The thickness of earth above the mine needed to be greater. A deeper mine could block almost all cosmic radiation but not, presumably, neutrinos.

Reines' venture into the East Rand mineshaft started with a ride that was part of the miners' daily routine. Every morning hundreds of workers awaited their turn to board the double-decker "cage" elevator. They were among more than 2000 miners who squeezed into a cage in the course of a day, twenty at a time on each of the cage's two decks. They stood pressed against one another as they began their descent. Almost all were black men along with an occasional white supervisor.

When Reines and his colleagues reached the end of the nearly mile-long initial descent, their ears were pounding. Discomfort from the changed pressure could usually be eased by yawning and swallowing. The

---

[2]D. Akimov, *et al.*, "Spotting a Ghost: Observation of Coherent Elastic Neutrino-Nucleus Scattering," *Science*, 367 (6536) September 15, 2017; Ann Finkbeiner, "Looking for Neutrinos, Nature's Ghost Particles," *Smithsonian Magazine*, November 2010, https://www.smithsonianmag.com/science-nature/looking-for-neutrinos-natures-ghost-particles-64200742/.

[3]"The Real Neutrino," *Time*, July 2, 1956.

Reines party rested and sipped tea as they acclimated to the increased underground pressure. An ear infection or even a common cold would have precluded a person from making the trip.

The East Rand mine was established in 1893, after deep veins of gold ore were discovered in the Witwatersrand region near Johannesburg. Excavation eventually deepened the mine to a half-mile, and then a mile. By mid-20th century it extended more than two miles down, becoming one of the deepest points on Earth that a human could reach. The deeper a mine, the greater the temperature and pressure. The temperature of the rock near the bottom exceeded 130 degrees Fahrenheit. Without a cooling system, human activity would be intolerable. Even with reduced temperature, conditions were clammy. As Fred and his colleagues continued their descent they were perspiring profusely. Sweaty discomfort was constant. Deep mine excavation also caused imbalanced pressure areas within the surrounding rock. As a result, spontaneous rock bursts could spew fragments with explosive force.

Reines' visit was facilitated by Professor Jacques Pierre Friederich (Friedel) Sellschop, director of nuclear physics research at the University of Witwatersrand in Johannesburg. Sellschop accompanied Reines and his colleague August (Gus) Hruschka through the labyrinthine passages to the bottom. The trip down also provided a view of the arduous labor conditions in the mine.

After the initial vertical drop, miners exited the elevator and headed to various work areas. Like most of the miners, Reines and his elevator colleagues walked half a mile to await another conveyance. A cart arrived on tracks angled at 45 degrees to take them farther down. At the end of that stretch, they walked to a different location to wait for another angled ride to the bottom. Completing the second mile down meant riding two different conveyances along some 4 miles of sloped tracks. The zigzag pathway had been created to follow the concentrations of the ore. The trip from the mine's surface to the bottom could take two hours or longer. By the time Reines reached the bottom, his clothing was soaked with perspiration.

Other than hard hats, the miners wore no protective gear. Most were garbed in ragged t-shirts and shorts; some were shirtless. The air was hot and muggy. The massive air conditioning system reduced the measured temperature to about 90 degrees. But water that was sprayed in the tunnels to keep the dust down produced 100 percent humidity. The temperature-humidity index — what it "feels like" — exceeded 110 degrees.

Years of excavation had produced a spider web of tunnels and passages at different levels of the mine. Throughout the branches, miners dragged themselves into four-foot-high alcoves. Their moist skin glistened under dim lamps. Once inside a "stope" a miner drilled holes in the surrounding ore-filled rock, packed them with dynamite, and moved from the vicinity. The area trembled from the subsequent explosions.

Miners later returned to the blast sites. With pickaxes and shovels they gathered loose pieces and loaded them onto carts. The packed carts rolled back up the inclined tracks to the elevator which lifted the fragments out of the mine. Upon reaching the surface the ore was moved to a processing plant and crushed to enable extraction of the gold. Meanwhile, the dynamite had made each stope several feet deeper. Continued blasting in one direction would eventually create a new tunnel. It also altered the pressure balance in the rock, which created the potential for rock bursts. In a year's time two dozen miners might be killed by an explosion, a rock fall, or a conveyance accident.

Beyond the dangers of the mine the political conditions of the country were a source of disquiet. South Africa was a land of natural beauty, but also of conflict. Its white rulers enforced a system of apartheid. Ironically, racism had become more entrenched there after World War II as African blacks elsewhere were shucking off European colonial rulers. The Nationalist Party, which had gained control of the government in 1948, vigorously promoted separation of the races. Blacks were prohibited from entering white residential areas without a permit. They were denied full participation in government, education, and other public affairs. The racist system did not end until 1992.

Fred Reines disdained apartheid. His attitude could be traced to a sentiment he expressed in his diary at age 16. On December 24, 1934, he read a newspaper story about a truce during World War I. The truce occurred in 1914, five months after the war began. On Christmas eve the sound of gunfire faded along the lines of combat. British, French, and Belgian soldiers on one side, and Germans on the other, sang Christmas carols to each other. The next morning troops emerged from the opposing trenches while calling out Merry Christmas to the other side. They gathered in "no man's land," shook hands, permitted the recovery of each side's dead soldiers, and exchanged food and souvenirs.

Fighting eventually resumed, but young Reines had been moved by the episode. "It's a shame that people haven't that fraternal feeling all the time," he wrote in his diary.[4] He had embraced the notion that harsh divisions could be mitigated if friendly attitudes were permitted to surface. Working across national boundaries could be a step toward enhancing fraternal feelings. Collaboration among scientists from different societies indeed might foster congenial behavior among them. At the same time, like many scientists, Reines believed that science should be "above" politics. He did not view his project as an accommodation to the racial policies of the regime. Rather, he would conduct it *despite* those policies in service to a purpose higher than politics, the advancement of scientific truth.

Sellschop shared these ideals. A white South African with a PhD in physics from England's Cambridge University, he was devoted to science. He was troubled both personally and professionally by his country's discriminatory policies. He valued international collaboration and was frustrated when scientists from other countries were discouraged by their governments from attending conferences in South Africa. At the time of Sellschop's death in 2002 his humanitarian efforts had been broadly recognized. Thabo Mbeki, who had succeeded Nelson Mandela as president of post-apartheid South Africa, paid tribute to his contributions. Mbeki presented a posthumous award to Sellschop recognizing his scientific achievements and his "love for freedom, racial harmony [and] equality."[5]

Despite the discomforts of working in the gold mine, Reines' visit convinced him of its suitability for neutrino- and other cosmic-related experiments. He was encouraged by the promise of support by Sellschop and the mine owners. His engineer friend Gus Hruschka believed that

---

[4] Roger Snodgrass, "Historical Society Embraces Reines' Nobel Legacy," *Los Alamos Monitor*, October 29, 2007, http://www.lamonitor.com/content/historical-society-embraces-reines-nobel-legacy.

[5] National Orders Under President Thabo Mbeki, South Africa, National Orders Booklet, December 10, 2002.

construction of a laboratory there was feasible. Working conditions for the research group, while posing safety risks, would be acceptable.

Reines was struck by the irony of traveling to one of the deepest points on Earth to detect particles streaming from outer space. He later wrote:

> It is an interesting commentary on the field of neutrino astronomy that we seek to see the heavens in a manner that does violence to common sense — we go deep underground. On reflection, however, it is obvious that the Earth is no barrier whatever to such neutrinos but serves to screen out the undesirable part of the cosmic radiation.[6]

In September 1963, an understanding about creating the subterranean laboratory was formalized. The mining company would provide the labor and resources to create the necessary space, facilitate access, and provide ongoing support for the project. Sellschop would arrange for the Reines research group to use his university's facilities. The group would run the new laboratory with Sellschop's participation. It would also construct the technical equipment at Case Institute for shipment to South Africa.

Before construction was underway, Reines was abruptly confronted with an unanticipated challenge. The legitimacy of the entire enterprise was threatened.

A *Reuters* news report dated September 30, 1963 was headlined "Atom Particle Will Be Sought: South Africa Goldmine Site of U.S. Test Next Year." The article began:

> A gold mine more than two miles down will be used in an American nuclear experiment. The experiment will take place at the East Rand Proprietary Mines at Boksburg, where a tunnel is to be driven 10,635 feet underground.

The text noted that Witwatersrand's Professor Sellschop "will cooperate with the Case Institute of Technology of Cleveland, in the experiment, scheduled to start next year." The article also dismissed any "suggestion that the experiment will involve the underground testing of a

---

[6]Frederick Reines, "Neutrino Astronomy," *Science*, October 1966; reprinted in W. Kropp, *et al.*, *Neutrinos and Other Matters*, Singapore: World Scientific, 1991: 318.

nuclear device." Sellschop was quoted as saying that Professor Fred Reines was the world's leading authority on the neutrino and that his project would help make "a reality of the dream of scientists to find out more about the neutrino."[7]

The apartheid regime was so distasteful to some critics that involvement of any sort with South Africa was deemed suspicious. Eduardo Mondlane, from Mozambique, was a leader of the African liberation movement. Six weeks after the *Reuters* story appeared, he declared before the United Nations Trustee Commission that the American and South African governments planned to conduct secret underground nuclear weapons testing. Khari R. Bagdelleh, a UN delegate from Tanganyika, echoed Mondlane with the claim that the two countries were "in agreement … with regard to testing underground bombs."

The allegations were alarming. Would the entire project be scrapped because of a false premise? Fred Reines immediately sought the help of colleagues close to American diplomatic authorities. An indignant Adlai Stevenson, the U.S. ambassador to the United Nations, publicly rejected the allegation with unusual vehemence. The claim that the project involved weapons research, he insisted, was "false, unfounded and irresponsible."[8]

Still, after Stevenson's statement, Mondlane reasserted his belief that the experiment "means more than finding out the reaction of cosmic rays in rocks."[9] The concern had been intensified by the confluence of three of the world's most contentious issues: the racial policies of the South African government, sensitivity about the spread of nuclear weapons, and resentment over perceived American interference in African affairs.

In time, with no apparent evidence for the nuclear weapons claim, the issue faded. But the attention given to the worrisome allegation underscored the speed with which misinformation could displace all other matters of international concern.

---

[7] "Atom Particle Will Be Sought: South Africa Goldmine Site of U.S. Test Next Year." *The New York Times*, October 1, 1963: 6.

[8] Sam Pope Brewer, "Atom-Blast Plan Is Denied by U.S.," *The New York Times*, November 14, 1964: 4.

[9] Howard Simons, "Tiny Neutrino Gives U.N. King-Size Headache," *The Washington Post*, November 16, 1963; reprinted in W. Kropp, *et al.*, "*Neutrinos and Other Matters*": 269–270.

Implementation of the plan for the laboratory began early in 1964. Mine workers started to blast away rock to create the space. At the same time, Reines and his Neutrino Group were designing and building equipment at Case for later shipment. During construction and afterwards, Reines' academic and administrative responsibilities kept him in Cleveland much of the time. He would visit the East Rand mine only sporadically. William Kropp had just received his PhD under Reines' supervision, and he eagerly accepted the offer to head the laboratory team at the mine in Fred's absence. Kropp and his family moved to Johannesburg in mid-1964 and remained there for nearly 3 years.

Prior to Kropp's arrival, Gus Hruschka was at the mine overseeing the project's development. After 6 months of blasting, the horizontal space for the laboratory was completed. It was 500 feet long. If stood on end it would have reached above the torch held high by the Statue of Liberty. Most of the space was in the form of a tunnel that was about 12 feet wide. The first of two key challenges to neutrino detection was now addressed. The laboratory site was so far underground that cosmic rays and other background radiation were largely blocked. The second challenge, the placement of detection apparatus, awaited its development at Case.

A year passed before the tons of laboratory equipment were produced, shipped 8000 miles to South Africa, and hauled overland to the mine. Lowering the fragile apparatus through the mine shaft took weeks of nerve-racking effort. Finally, after arriving deep in the mine, the materials could be sorted and assembled. The new laboratory included a wooden structure, which was referred to as the hut. It was built onsite to hold cables, vacuum tubes, monitoring screens, and other electronic paraphernalia. The hut was air conditioned and served as the main working area for the scientists and technicians.

The laboratory was not immune to danger. On one occasion, sections of the detector were crushed by rock bursts and had to be replaced. Another time a nearby dynamite explosion caused the hut to collapse. Fortunately, nobody was inside, but the setback required rebuilding and restoring the wood structure with its panoply of electronic gear.

When Reines was in the United States, he remained in frequent contact with Kropp and the other lab workers. They communicated mostly by teletype and occasionally by telephone. Reines kept up with their progress and their challenges. He advised, encouraged, and continued to help steer the research efforts.

The resident crew of a half-dozen scientists and technicians spent much of their time in the laboratory hut. At first their schedule was especially trying. Days in the mine alternated with days at the University of Witwatersrand to analyze the accumulating data. A workday began with a member of the crew getting up at 5 a.m. and driving to pick up the other members from their Johannesburg homes. They would arrive at the mine by 7, change into t-shirts and jeans, don hard hats, and squeeze into the elevator cages with the miners. The crew was occupied with keeping the elaborate detector system in operating order. Their work often continued into the night and they might not get home until after midnight.

Decades later I recapitulated to Kropp the experience that he and others had described about their work in the gold mine. The heat and humidity, the danger from rock bursts and poison gas, the long hours in the laboratory, the uncomfortable morning ride to the bottom and the late-night return to the top. All with uncertainty that their experiments would yield satisfying results, I wondered.

"Well that's what an experiment is about," Kropp replied. "We were young and crazy then; it was fun and we were doing physics and learning things."[10] In 1966, two years after Kropp moved to Johannesburg, Henry Sobel, another of Reines' graduate students, arrived to join the research group. Sobel became one of the group's leaders after Kropp returned to the United States. He remained in South Africa until 1969. Soon afterwards, work at the gold mine laboratory began to phase out.

When I asked Sobel the same question about danger that I had asked Kropp, his answer was similar. "We were young and stupid, so we didn't realize how dangerous it was." He added, "Yes, but it was exciting,"[11]

Kropp conceded that in retrospect the team was lucky that no one in the group was ever seriously hurt. Like Sobel, he couldn't resist reaffirming the excitement that drew them and Fred Reines and legions of other explorers to embark on daunting, even hazardous, ventures.

The gold mine project was a big Reines idea. Whether it would be successful remained uncertain, pending the findings of the experiment

---

[10] William (Bill) Kropp interviews, 2017–2019.

[11] Henry (Hank) Sobel interviews, 2017–2019.

(reported in Chapter 10). Meanwhile, Fred's determination was consistent with a charming characterization by John Wheeler, the eminent 20th century physicist who coined the term "black hole." Reines, he said, was "talented in both theory and experiment, a bear of a man given to thinking big about nearly impossible problems as he paced up and down in his oversize shoes."[12]

In 1960, I met John Wheeler. Based at Princeton University, he had come for a meeting at the University of California, Berkeley, where I was studying. He was soft spoken and unassuming. He mentioned Fred Reines and their many conversations at Los Alamos and at Eniwetok, a South Pacific island where nuclear testing had taken place. He extolled Reines' big ideas.

Reines' devotion to science and regard for individuals who excelled in science had roots in his childhood experiences. They seeded his life-long quest for knowledge, for accomplishment, and for recognition.

Fred Reines, Gus Hruschka prepared to enter the East Rand Gold Mine, South Africa. Reines Collection.

[12] John Archibald Wheeler with Kenneth W. Ford, *Geons, Black Holes, and Quantum Foam: A Life in Physics*. NY: W.W. Norton, 2000: 281.

Research Team on the way to the neutrino detector at the bottom of the East Rand Gold Mine. Peter Holz, "In Search of Neutrinos," *South African Panorama, July 1968*. Reines Collection.

Chapter 2

Language, Science, and the Ghost Particle

What is a muon? After a contemplative pause a friend answered my question by asking, Is it like a moron? Not really, I chuckled. I posed the same question to many non-physicist friends and colleagues. Most responses were equally clueless, if less entertaining.

A few knew that it was a subatomic particle of some sort. No one knew that it was negatively charged like an electron, but heavier. Created by cosmic ray collisions with particles in the earth's upper atmosphere, a muon is unstable. In about two millionths of a second (long enough for it to reach Earth's surface) it "decays" into an electron and two neutrinos. The term neutrino is somewhat more familiar to the public. Improbable powers were attributed to the particle in television's science fiction series, *Star Trek*. It also received mention by actors playing nerdy physicists in the long-running popular sitcom, *The Big Bang Theory*.

Neutrinos come in different forms, sometimes referred to as flavors. Antineutrinos also exist, which are particles of antimatter that correspond to those in ordinary matter. The differences between matter and antimatter particles are based partly on their opposite electrical charges. An electron has a negative charge and its mirrored partner, the positron (an antielectron), has a positive charge. If two such particles collide, they annihilate each other and become photons — particles of light. Photons can carry energy despite having no mass and no charge. Since a neutrino, like a photon, has no electrical charge, scientists are contending with the seemingly bizarre possibility that a neutrino and an antineutrino may be the same thing.

Equal quantities of matter and antimatter at the time of the Big Bang would have precluded creation of the universe as we know it. They would have destroyed each other. The predominance of matter rather than antimatter is believed traceable to conditions at that early time. The neutrino is thought to have been central to the outcome. In this book, reference to a neutrino includes all forms of the particle insofar as they share fundamentally common properties.

The vocabulary of physics can be challenging to a non-scientist. Take the word "decay." Its everyday meaning connotes irreversible deterioration, usually bacterial destruction of an organic substance. Apples decay and turn rotten; teeth do so in the form of cavities. But to a physicist or chemist decay may mean something else.

Radioactive decay refers to the natural emission of particles from certain atomic nuclei. The consequence may be creation of a totally different element. The loss of a so-called "alpha particle" — a combination of two protons and two neutrons — from a uranium atom makes it the nucleus of a thorium atom. No deterioration — just transformation into a different and slightly lighter element.

A neutron may undergo transformation into an electron, proton, and a neutrino (actually an antineutrino). This is called beta decay because the electron is also known as a beta particle. A free neutron decays in just this way, with a half-life of about 10 minutes. Within a nucleus, a neutron is stabilized. Around the turn of the 20th century three kinds of radiation from radioactive atoms were identified — alpha, beta, and gamma.

To add to the notational overload, there is a process called "inverse beta decay." In this process, a neutrino (antineutrino) reacts with a proton to form a positron (antielectron) and a neutron. This is the process that Fred Reines and Clyde Cowan utilized when they were the first to detect the neutrino (actually understood to be an antineutrino).

In 2006, Herald Kruse, a veteran of the Reines–Cowan team that detected the neutrino in 1956, offered a retrospective that included the notation: $^{-}\nu + p \rightarrow \beta^{+} + n$ ($^{-}\nu$ = antineutrino; p = proton; β^{+} = positron; n = neutron). He then reflected on the uncertainty at that time about the validity of the formulation:

> I have seen Reines write this reaction on a blackboard countless times in various discussions, and it is so familiar now that I fear it has become

> prosaic. But in 1956 this reaction was hypothetical and to our team it presented an enormous challenge. We were attempting to establish the reaction either as a reality or a figment of a "desperate" imagination, both possibilities with far reaching results.[1]

Another term with distinctive meaning for physicists is cross section. To most people a cross section is a surface exposed by a transverse slice through an object. To particle physicists the meaning is very different. It describes the probability of two particles interacting under certain conditions: including the number of particles in the beam and the angle at which they hit a nuclear target, analogous to the circular target at which arrows are aimed.

A cross section relates not just to the probability that one particle will interact with another, but that it will do so in a particular way. Thus, there is a specific cross section for inverse beta decay, measuring the likelihood that an antineutrino will react with a proton to create a positron and a neutron. This particular cross section is exceedingly small. As neutrinos race through the universe an almost unbelievably tiny fraction of them will eventually interact with a proton. The cross section for such a reaction is indeed minuscule.[2]

Long after his discovery of the neutrino, Reines continued to chase the elusive particle. He was steeped in efforts to identify the neutrino's varying characteristics, its earthly and cosmic sources, and its importance in the workings of the universe. An information sheet at Fermilab near Chicago includes several facts about neutrinos:

- They are elementary: "Neutrinos are fundamental particles, which means that — like quarks and photons and electrons — they cannot be broken down into any smaller bits."
- They are abundant: "Roughly a thousand trillion of them pass harmlessly through your body every second."

---

[1] Herald W. Kruse, "Reines-Cowan Team Discovery of the Electron Neutrino," *Nuclear Physics B* (Proc. Suppl.) 2011.

[2] This information derives from *Symmetry* magazine, https://www.symmetrymagazine.org/article/speak-physics-what-is-a-cross-section and an email communication from physicist Kenneth Ford, 2019.

- They weigh almost nothing: "Neutrino masses are so small that so far no experiment has succeeded in measuring them."
- As is now understood, their sources are many and diverse: "Neutrinos are created in many processes in nature. They are produced in the nuclear reactions in the sun, particle decays in the Earth, and the explosions of stars. They are also produced by particle accelerators and in nuclear power plants."[3]

These current understandings underscore the immense gap in knowledge that has been narrowed since neutrinos were first detected. At that time, the sole detected source was a nuclear reactor that produced low energy neutrinos.

Reines' creativity and dedication inspired many other scientists to engage in neutrino research. Only some of his significant findings about the neutrino are recapitulated here. They are embedded in the tale of a man whose deepest passion was science, but whose talents and interests extended beyond. They ranged from singing and acting to composing essays and poetry. He would employ these features in his ardor to share with others his love of science.

Fred's daughter, Lisa (Alisa Reines Cowden), recalled that her father would, unprompted, suddenly break into a musical piece by voice or whistle. Yet, she added, one could not understand what it was like to be around him unless you appreciated his passion for science.

> He was a physicist all the time. Everybody has their normal in their households and that was normal to us. That's just the way he was. He was fascinated about his work. When he would discuss his work I often didn't understand it, but I could see he was fascinated by it.[4]

In 1966, Reines became founding Dean of the School of Physical Sciences at the University of California at Irvine. There he taught a physics course for non-science majors. He once responded to an inquiry about the course from a fellow physicist. His aim, he wrote, was to show

---

[3] The characteristics are listed in *Symmetry*, https://neutrinos.fnal.gov/wp-content/uploads/2018/04/NeutrinoArePoster_Final_v2.pdf.

[4] Alisa (Lisa) Reines Cowden interview, 2017.

the students that "science is not a thing apart from poetry, philosophy, the arts." Rather, "it represents but one facet of man's creativity, but [in] a distinctive manner to ask questions of nature."[5] Experimental demonstrations in his classroom included "only the simplest equipment," he wrote.

> I use a candle, a piece of cardboard with a hole in it, a large lens, a piece of string, a paper clip. I whistle along with a student to illustrate beat frequencies.

Absent from his course were complex equations and formulas, an approach to popular engagement that I seek as well in this book.

In 1930, Austrian-born Wolfgang Pauli was contemplating an issue that troubled many in the physics community. It had to do with the "decay" of neutrons in the nuclei of radioactive atoms. A neutron, which carries no electric charge, is converted into an electron (a negatively charged particle) and a proton (a positively charged particle). Some energy appeared to vanish inexplicably in the process. The phenomenon seemed to violate the law of energy conservation, which holds that energy cannot be created or destroyed, but only change in form. Or, as another physicist has observed: "what goes in must equal what comes out."[6] In this case, what was coming out seemed to be less than what was going in.

To resolve the dilemma Pauli proposed the existence of a yet-unseen minuscule particle with no electric charge and very little or no mass. His proposition came with a disturbing caveat. In a letter to colleagues he wrote, "I have done something very bad today by proposing a particle that cannot be detected."[7]

In 1934, with detectability still doubted by most scientists, Enrico Fermi expanded on Pauli's idea. An electron is also known as a beta particle and is emitted in "beta decay." Fermi called Pauli's postulated

---

[5] Letter to Dr. M. J. Renan, Nuclear Physics Research Unit, University of Witwatersrand, July 5, 1972, Reines Collection.

[6] Jayawardhana: 7.

[7] American Physical Society, Looking at Physics History, https://www.aps.org/publications/capitolhillquarterly/201110/physicshistory.cfm.

particle a "neutrino." Italian for "little neutral one," the name was first suggested to Fermi by a colleague, Edoardo Amaldi.[8]

Enrico Fermi. Courtesy of the Atomic Heritage Foundation.

Two decades later, Fred Reines and Clyde Cowan devised an experiment that they believed could validate or negate the existence of the particle. Their experiment confirmed its detection in 1956. Subsequently Fred continued his research with other collaborators, as did Clyde until his death in 1974 at age 54. Since the prize is not awarded posthumously, the Nobel committee eventually recognized only Fred, although in fact he was awarded half of the physics prize. The other half went to Martin Perl of Stanford University for his discovery of the tau, another elementary particle in the same general family as the electron and the neutrino. The committee's 1995 announcement lauded Fred not only for detecting the neutrino, but also for his later findings that advanced the field of neutrino physics.

The elusive nature of the neutrino had prompted Fred to characterize it as ghostly. He and Clyde Cowan described their 1950s efforts as Project Poltergeist, referring to a presumed ghost that causes strange noises and

---

[8]Edoardo Amaldi, *Physical Reports*, Vol. 111, 1984; reference no. 277, http://www-numi.fnal.gov/public/story.html.

movements of physical objects. The description seemed apt — an apparently mysterious entity responsible for certain physical behaviors.

Even though the neutrino's existence has been validated, the supernatural metaphor persists. Many trillions of neutrinos might pass through a detector unnoticed before one by chance interacts with another particle leaving a tell-tail signal that the ghost had been present. Appreciation of the importance of the neutrino has grown with time, and accordingly so has the significance of Fred's confirmatory experiments. Thus far the physics Nobel Prize has been awarded four times for findings about the neutrino. Eight individuals have been recipients, some of whom worked together as a team and others whose separate investigations led to parallel discoveries. The first neutrino-related Nobel went to a group in 1988. Leon Lederman, Melvin Schwartz, and Jack Steinberger had developed a method of detecting a second form, or "flavor," of the elusive particle. Fred's earlier discovery preceded recognition that the particle could appear in more than one flavor.

In 2002, Raymond Davis from the United States and Masatoshi Koshiba from Japan, shared the prize. Each had detected neutrinos coming from the sun. Davis' experiments had been conducted with theoretical advice from John Bahcall in the 1960s. Hydrogen is the most abundant element in the universe, and the constituents of hydrogen atoms — protons and electrons — are plentiful in the core of the sun. The sun's immense gravity and associated high temperature can force the protons to fuse, becoming the nuclei of helium atoms (alpha particles). The fusion process generates neutrinos as it emits an enormous amount of energy.

For thousands of years people ascribed supernatural attributes to the sun and its energy, and societies around the globe viewed the sun as a deity. The ancient Egyptians believed the sun was "the all-father of creation." Native American tribes performed sun-dances as acts of propitiation. Until the end of World War II, every emperor of Japan was deemed a direct descendant of the sun-goddess Amaterasu.

The sun's source of light and heat continued to puzzle non-fabulist scientists into the 20th century. By the 1930s, nuclear fusion had become commonly viewed as a key to the process. The Davis-Bahcall finding that neutrinos were continuously streaming from the sun's core validated the proposition that nuclear fusion was the source of the sun's energy.

Many of Bahcall's peers believe that he should have been included among the Nobel winners. He died in 2005. In tacit appreciation of his work the Nobel organization's website continues to include essays by Bahcall, including one titled, "How the Sun Shines." It places the neutrino at the center of a phenomenon that has caused wonderment throughout human history.[9]

More recently, in 2015, the Nobel Prize went to Canada's Arthur McDonald and Japan's Takaaki Kajita. Toward the end of the 20th century, each had led experiments demonstrating that neutrinos had mass. Until then many scientists believed that neutrinos, like photons — particles of light — lacked any mass.

The notion that neutrinos were massless was abetted by an apparent lack of evidence to the contrary. Years earlier when I'd ask Fred what he thought about the question, he'd reply with a shrug that "the jury is still out." But he was well aware of the importance of the debate and the cosmic implications of its resolution.

The significance of this seemingly benign issue was made clear by the Nobel selection committee in 2015. McDonald and Kajita's experiments confirmed that neutrinos "oscillate," which means they have mass. Oscillations are due to a neutrino's mix of three flavors, each of which may travel and interact with definite masses. The finding that neutrinos change identities, the committee said, "requires that neutrinos have mass [which] changed our understanding of the innermost workings of matter."

In a popularized version of the finding, astrophysicist Stuart Clark wrote a *Guardian* article titled: "How Neutrinos Saved the World." How so? Life on Earth could not exist without the heat and light produced by the sun. Confirmation that neutrinos had mass broadened understanding of energy production in the core of the sun and other stars. Estimates now hold that the sun will continue to "burn" for another 5 billion years.[10]

---

[9] John N. Bahcall, "How the Sun Shines," The Nobel Prize, http://www.nobelprize.org/nobel_prizes/themes/physics/fusion/. Also, Bahcall, "Solving the Mystery of the Missing Neutrinos," http://www.nobelprize.org/nobel_prizes/themes/physics/bahcall/.

[10] Stuart Clark, "How Neutrinos Saved the World," *The Guardian*, October 15, 2015, https://www.theguardian.com/science/across-the-universe/2015/oct/06/nobel-prize-for-physics-2015-how-neutrinos-saved-the-world.

If neutrinos had proved massless, the sun's energy would be depleted much sooner. The consequence would mean a frozen, lifeless earth in some hundreds of thousands of years, not billions, according to the article.

Many scientists considered the proposition unlikely, even silly. Clark himself revealed a playful skepticism at the conclusion of his article: "And by proving that there was nothing wrong with the nuclear reactions inside the sun, they [Kajita and McDonald] saved the world from an icy death (sort of)."

A more imminent if less cataclysmic effect was revealed in a leading scientific journal. A 2015 article in *Science* magazine was titled: "How Neutrinos Saved Your Teeth from Cavities." Fluoride compounds in drinking water and toothpaste sharply reduce the incidence of tooth decay. In the core of stars, where elements are formed, fluorine is known to be much scarcer than oxygen and neon despite their similar atomic structures. An atom of oxygen contains 8 protons, an atom of fluorine 9, and of neon 10.

When a star explodes it becomes a supernova, spewing massive quantities of neutrinos and other particles into the universe. In the process, an occasional neutrino interaction may abet a proton's loss from a neon atom. Or, conversely, it may induce the addition of a proton to an oxygen atom. Either action transforms those atoms into fluorine. The transformation thus expands the availability of this dental decay-fighting element.

Verne Smith, an astronomer at the National Optical Astronomy Observatory in Tucson, Arizona, suspects that supernova neutrinos account for as much as two-thirds of Earth's fluorine. Saving teeth may be of less cosmic significance than saving the world. But the variety of proposed effects by the ghost particle continues to grow and beguile.[11]

Heinrich Paes, a neutrino expert at Dortmund University in Germany, has written that the point of neutrino physics is to gain "a better basic understanding of the universe [and further a] rational world view." But he also noted that scientists have discussed neutrinos as potentially useful for applied purposes. According to some, intense neutrino beams might be used to destroy nuclear weapons. Others have suggested that neutrino beams could enhance messaging for submerged modern submarines.

---

[11] Ken Croswell, "How Neutrinos Saved Your Teeth from Cavities," *Science*, September 6, 2015, http://www.sciencemag.org/news/2015/09/how-neutrinos-saved-your-teeth-cavities.

Currently communicating in deep water is limited and submarines must surface to use high bandwidths. With neutrino communication, submarines will be able to remain under water for longer periods. Still others have suggested that neutrino beams might enhance efforts to contact civilizations in distant stellar systems.[12]

One neutrino project that is already under development aims to target countries engaged in illicit nuclear weapons programs. A detector in the vicinity of a nuclear reactor might identify the number of antineutrinos emitted by the reactor. This in turn would reveal whether its volume of uranium and plutonium is consistent with peaceful energy production or with development of bombs. Completion of the test device is expected by 2023. Adam Bernstein, a physicist at the Lawrence Livermore National Laboratory in California, is leading the effort, which he describes as a "nice duality between basic science research and nonproliferation."[13]

These observations and every other consequential finding about the neutrino came after Reines and Cowan's discovery. Physicist Francis Halzen noted that prior to their discovery many people believed the presumed particle "was like a mathematical trick to make the theory work."[14] Their discovery was the indispensable first experimental step. As the importance of the neutrino became increasingly evident, more and more scientists gravitated to its study. By the time Fred died in 1998, three years after receiving the Nobel, he was regarded as "the father of neutrino physics" and his neutrino experiments had "set in motion a new way of looking at the universe."[15] Neutrino researchers at the time of Fred and Clyde's 1956 discovery numbered about a dozen including several who

---

[12] Heinrich Paes, *The Perfect Wave: With Neutrinos at the Boundary of Space and Time*. Cambridge, MA: Harvard University Press, 2014, Chapter 16.

[13] Kenneth Chang, "How to Spot a Nuclear Bomb Program? Look for Ghostly Particles," *The New York Times*, March 27, 2018, https://www.nytimes.com/2018/03/27/science/nuclear-bombs-antineutrinos.html.

[14] Francis Halzen, World Science Festival, 2012, https://www.youtube.com/watch?v=CBfUHzkcaHQ.

[15] John Noble Wilford, "Frederick Reines Dies at 80; Nobelist Discovered Neutrino," *The New York Times*, August 28, 1998; Jeff Gottlieb, "Frederick Reines: UC Irvine Professor Discovered Neutrinos," *Los Angeles Times*, August 28, 1998, https://www.latimes.com/archives/la-xpm-1998-aug-28-mn-17496-story.html.

had been part of their research team. Ever since, the number, worldwide, has continued to grow. More than 400 investigators co-authored a recent report on the detection of the varied forms of the particle.[16]

Physicist William Kropp recalled that when he began working with Fred in the early 1960s, there were perhaps 20 neutrino scientists in the world. Ray Jayawardhana, the author of *Neutrino Hunters,* estimates that the number of neutrino investigators now exceeds one thousand.[17] Other scientists I interviewed believe the figure is closer to 3000. A U.S. Energy Department panel of experts recently concluded that the study of neutrinos offered the most promise for continued discovery in all of particle physics.[18]

Since Reines' discovery, fascination with the particle has been expressed both by scientists and nonscientists. Enthrallment with the ghost particle has also found a perch in fiction and poetry. Episodes of *Star Trek*, television's longest running science fiction series, are replete with references to neutrinos. They appear in improbable scenarios to disguise the presence of a spaceship, to detect space weapons, to kill subatomic-sized bacteria.

In 1960, 4 years after the Reines–Cowan detection, John Updike was inspired to write a poem about the increasingly famous particle. At the time, many physicists believed the neutrino lacked mass. That assumption, later overturned, was reflected in Updike's poem.

Cosmic Gall

Neutrinos, they are very small.
They have no charge and have no mass
And do not interact at all.
The earth is just a silly ball
To them, through which they simply pass,

[16] K. Abe, *et al.*, "Indication of Electron Neutrino Appearance from an Accelerator-produced Off-axis Muon Neutrino Beam," Preprint submitted to PRL, https://arxiv.org/pdf/1106.2822.pdf.

[17] Jayawardhana: 171.

[18] Jessica Orwig, "Neutrinos top list of targets for US particle physics," *New Scientist*, May 22, 2014, https://www.newscientist.com/article/dn25612-neutrinos-top-list-of-targets-for-us-particle-physics/.

Like dustmaids down a drafty hall
Or photons through a sheet of glass.
They snub the most exquisite gas,
Ignore the most substantial wall,
Cold-shoulder steel and sounding brass,
Insult the stallion in his stall,
And, scorning barriers of class,
Infiltrate you and me! Like tall
And painless guillotines, they fall
Down through our heads into the grass.
At night, they enter at Nepal
And pierce the lover and his lass
From underneath the bed — you call
It wonderful; I call it crass.

Nearly 60 years later, in 2018, the neutrino prompted a question on the website *Quora*: "Why is the neutrino the most important particle in nuclear physics?" The site, founded by former employees of Facebook, provides a question-and-answer platform on various subjects. It has drawn millions of participants. Responses by scientists to the question varied. They ranged from denial that any one particle was "the" most important, to the neutrino's centrality in unraveling cosmic mysteries.[19]

In fact, a chapter in Heinrich Paes' 2014 book about cosmology and the neutrino was titled, *Neutrinos: Key to the Universe*. Paes recounts neutrino experiments, including some on which he collaborated, that led to a conclusion that he called mind-boggling: "According to the ideas outlined in this chapter, the neutrino could be the reason we exist — indeed it may even be the cause of the existence of the entire universe."[20]

The most abundant of all particles with any mass, neutrinos are part of the process that powers the sun and stars and that creates elements including iron, nickel, and heavier metals. Neutrinos produced by the Big Bang, 13.8 billion years ago, continue to travel throughout space, many at nearly the speed of light. As they pass through the Earth and our

[19] *Quora*, https://www.quora.com/Why-is-the-neutrino-the-most-important-particle-in-nuclear-physics; Fermilab, "Why Study Neutrinos?" http://minerva.fnal.gov/why-study-neutrinos/.
[20] Paes: Chapter 11.

underground neutrino telescopes, they afford a potential to reveal more secrets about the birth of the universe.

An internet search in 2020 identified more than 50 neutrino-experiment locations including some under construction.[21] Reines' own description of the particle is Haiku-like — brief, simple, thoughtful. He called the neutrino "the most tiny quantity of reality ever imagined by a human being." It was Fred Reines who became the human being that transformed the imagined particle into reality.

[21] List of Neutrino Experiments, Wikipedia, https://en.wikipedia.org/wiki/List_of_neutrino_experiments.

Part II

Becoming

Chapter 3

Becoming a Physicist

My aunt Gussie met and married Israel Reines in New York City in the early 1900s. They each had emigrated from the Bialystok region, then part of Russia and now Poland. Gussie's family had lived in the small Jewish village of Krinik. After enduring several pogroms — violent group attacks against Jews — her family emigrated to the United States. Upon arrival, her father (my grandfather) changed the family name from Krinsky to Cohen.

Mr. Reines was a strong figure. Manually skilled, he became a silk mill operator in Paterson, New Jersey. The city's silk industry had flourished since the 19th century and he owned a mill for a brief period after World War I. But when silk manufacturing began shifting to southern states, Paterson's industry dwindled and Reines turned to running a grocery store.

Israel Reines also brought to the marriage a degree of pride by association. His uncle, Isaac Jacob Reines, had been a famous Russian rabbi at the turn of the century. Rabbi Reines advocated teaching secular subjects along with religion in Orthodox Jewish schools. He was also the founder of Mizrachi, a movement that supported reestablishment of a Jewish national homeland in Israel. Major streets in Tel Aviv and Jerusalem are named after the great Rabbi Reines.

Aunt Gussie was the oldest of seven siblings, and the only female. My father, Morris, was the youngest, and he especially looked up to his sister. Among my earliest memories was my father's visible warmth toward Gussie and Israel's four children, the youngest of whom was Fred.

Fred Reines was born on March 16, 1918, in Paterson, New Jersey. Three years later his family moved to Hillburn, New York, where his parents ran a grocery store. The family, including Fred's older siblings — a sister, Paula, and brothers William and David — lived above the store. The modest two-story building still stands on a 50-by-110-foot lot at 36 Mountain Avenue. When Fred was 8, the 5-year lease on the building expired and the family returned to New Jersey. A lifetime later, at his Nobel ceremony, it was clear from Reines' lecture that recollections of Hillburn remained ingrained: "My early childhood memories center around this typical American country store and life in a small American town, including 4th of July celebrations marked by fireworks and patriotic music played from a pavilion bandstand."

The Reines children attended Hillburn's Main School, a block from the store. The school, later renamed the Hillburn School, has become the administrative building for the Suffern Central School District. In 2015, the former school was listed in the National Register of Historic Places for its role in the 20th century civil rights struggle. A memorial plaque stands near the entryway:

> "On this site in 1943, Thurgood Marshall, NAACP Legal Director and later US Supreme Court Justice, led the movement to end racial segregation in Hillburn schools."

The Main School, a sturdy cement structure whose students were white, contained a library, playground, and indoor plumbing. Black students attended the smaller, wooden Brook School, which had no amenities. Marshall's petition prompted the New York State education commissioner to close the Brook School and send all of Hillburn's students to the Main School. The action anticipated the landmark 1954 Supreme Court desegregation decision, *Brown v. Board of Education*.[1]

Calls to desegregate the Hillburn schools, which came largely from parents of the black children, were initially resisted by many in the white community. But when the state commissioner closed the Brook School, its students were transferred to the Main School without incident. The Main School became a model of successful integration. The Reines family had long since moved away and Fred almost certainly was unaware of the salutary role of his primary grade school in the civil rights movement. As an adult his embrace of equal rights for all, regardless of race, ethnicity, or religion, was clear. Had he known about his primary school's role he likely would have mentioned it in his Nobel remarks.

In 1926, back in Paterson, the Reines family was living on the second floor of a two-family house at 101 12th Avenue. On the first floor were Fred's (and my) grandparents, Philip and Rebecca Cohen, who owned the house. Both apartments were modest in size and decoration, though each was filled with books. Fred's parents lacked formal education, though they were keenly intelligent. A family member once said he might be spending a costly sum to expand his tiny office. He wondered whether to continue renting or try to buy the overpriced property. "You have a choice?" Aunt Gussie asked. "Yes." "OK, then buy. Don't sew your buttons on another man's vest." Her advice struck me as sound, and her metaphor clever.

In 1928, when Fred was nine years old, he posed for a photograph in front of a tree in the neighborhood. He was flanked by grinning playmates whose feet were planted on the ground. But Fred's legs were wrapped around a sturdy branch above their heads and he was hanging upside down between them. His expression was one of utter nonchalance. Besides demonstrating physical agility, his pose suggested a characteristic that would manifest in later years as well. He was a performer, liked to stand

[1] Jane Lerner, "Thurgood Marshall's Hillburn Desegregation Battle Recalled," *USA Today*, March 29, 2016, https://www.usatoday.com/story/news/nation-now/2016/03/29/thurgood-marshall-civil-rights-battle-hillburn-school/82401894/.

out, and unabashedly veered from the conventional manner of doing things.

Fred suspended upside down from a tree, age 9. Reines Collection.

Fred recalled that as a child, he enjoyed building things and participating in group singing at school. Music and singing remained lifelong interests. His initial attraction to science came unexpectedly, when he witnessed a curious bending of light as it passed through a tiny opening.

> [It] occurred during a moment of boredom at religious school, when, looking out of the window at twilight through a hand curled to simulate a telescope, I noticed something peculiar about the light; it was the phenomenon of diffraction. That began for me a fascination with light.[2]

---

[2] Frederick Reines, The Nobel Prize Biographical, https://www.nobelprize.org/prizes/physics/1995/reines/biographical/.

Fred, age 10. Reines Collection.

Fred's brother David, two years his senior, was, like Fred, athletically inclined. David later became a skilled tennis player. As Fred entered his teens, he and David practiced handstands and other gymnastic moves together. Working out with his brother became the forerunner of Fred's gymnastic proficiency during his high school and college years.

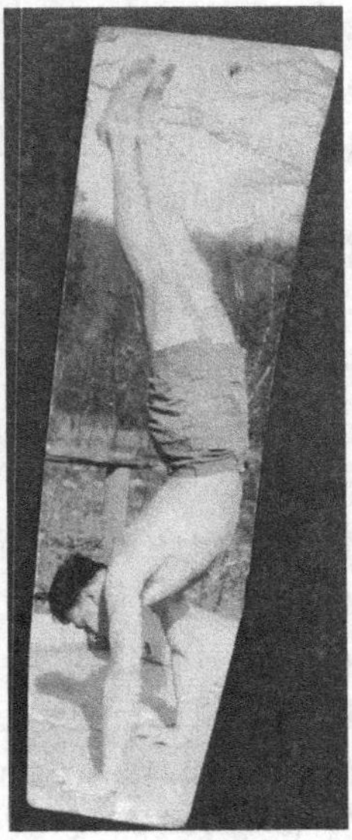

Fred, age 15. Reines Collection.

Educational achievement was a preeminent value for the Reines children. In Paterson's Eastside High School *Class Book*, June 1929, the two oldest, William and Paula, appear on p. 48. (They graduated together even though William was a year older. He had begun first grade late and she had skipped a half-year in elementary school. At that time in many states an elementary or high school year was divided into separate fall and spring sessions. Graduations were held twice a year, in January and June.)

Both were at the top of the class. Classmates admiringly described William as dignified and Paula as certain to realize her ambitions. Their records were remarkably similar. William was editor-in-chief of the *Class Book*, and Paula editor-in-chief of the *Criterion*, the school newspaper. Each of them authored a full-page essay in the *Class Book*, Paula on the Class History and William on the Class Will.

Despite their parallel achievements they bore contrasting dispositions. Paula was outwardly competitive and William casual and unflappable. David, their younger brother, was somewhat shy, though he too was intellectually accomplished. William and David became lawyers and Paula a physician. Whatever the difference in personalities among the three older siblings, they all shared strong affection for their baby brother. Fred's later scientific achievements only enhanced their admiration.

A household environment in which erudition and accomplishment were so highly prized might have intimidated some youngsters, but not Fred. Soon after his oldest two siblings graduated from high school, the family moved to nearby North Bergen, where Fred completed grammar school. Because the town lacked a high school, Fred attended one in neighboring Union City. Fred's grades in science initially trailed those in literary studies. By his junior year the order of his academic accomplishment had reversed, in part due to an inspirational science teacher. He graduated from Union Hill High School in 1935. The list of activities next to his yearbook photograph was as eclectic as those of William and Paula seven years earlier: editor-in-chief of his school's yearbook, president of the History Forum and a member of the Language Club, the Dramatic Club, and the Student Council. In addition, he was a leader on the Gym Team. His stated ambition was, "To be a physicist extraordinary."

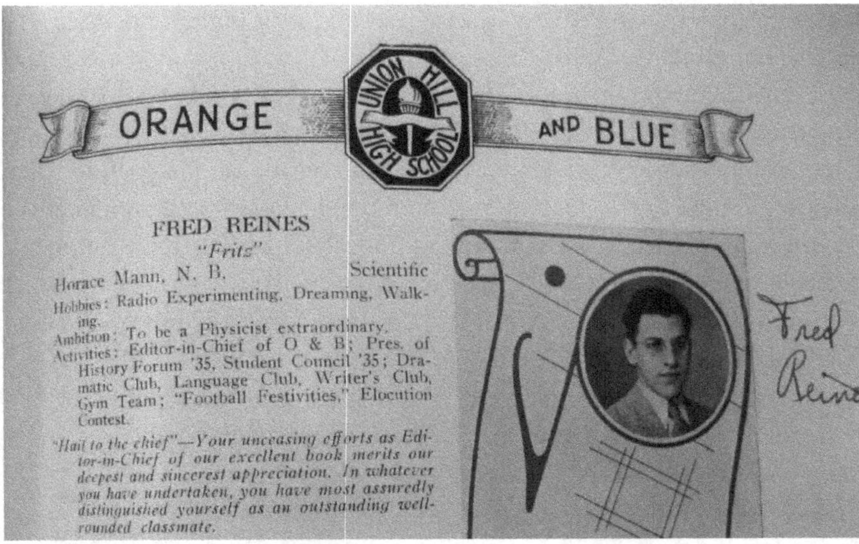

1935 High School Yearbook.

Fred's interest in music was evinced both as listener and performer. One time, after dinner at his parents' apartment, someone asked him to sing. Unhesitatingly he delivered an *a cappella* medley from George Gershwin's "Porgy and Bess" and we were all transfixed by his rendition of "Summertime." His roll out of "and the 'livin' is easy" bounced off the dining room walls.

Fred was a forceful figure. Tall and strong, he had become an Eagle Scout while in high school, which required not only leadership but also skills to acquire numerous merit badges for athletic, camping, and academic proficiency. Fewer than four percent of the millions of boys in the scouting movement attain the rank of Eagle.

He received bachelor's and master's degrees at the Stevens Institute of Technology in New Jersey, then an all-male school. While studying engineering and science, he was also drawn to music and briefly considered a career in singing and theater. He had even taken lessons from a voice coach at the Metropolitan Opera. But his attraction to music ultimately did not displace his deeper love for science. At age 20 and a third-year college student, he confided in his diary his ultimate and prophetic ambition: "I'd like to discover one little thing and build my career around it."[3]

Fred's photograph in the Stevens Institute yearbook, class of 1939, projects an air of self-assurance. With the hint of a smile he peers straight ahead as if seeking eye contact with the viewer of his picture. The description next to the photograph noted that he could be argumentative. But the stunning list of his activities is no less impressive than the one in his high school yearbook. In the course of his undergraduate years at Stevens he had been a member of the Glee Club, Dramatic Society, Banquet Committee, Numerals Club, and Calculus "Cremation" Committee. He was managing editor of *The Link* — the school's annual yearbook — and president of the Stevens Engineering Society.

FREDERICK REINES

FRED

EVER since September, 1935, Fred has managed to keep himself in the public eye, the public in this case being '39. His powerful physique and his skill on the gym apparatus started things. This same physique stood him in good stead while he twice became '39's Heavyweight Cane Spree Champion.

Our eyes lit up at his various histrionic ventures for the D. S. and, in particular, his immortal characterization of "Doc" Pond at our Freshman Banquet. At other times, such as the unveiling of one of his less humorous jokes, or during one of his fruitless and time-consuming arguments with the profs, we have been forced to wipe Fred out of our eye much as a squirt of grapefruit. The squirts are diminishing though, so we forgive him.

LINK (2, 3); Managing Editor (3); Glee Club (3); Secretary-Treasurer (3); Dramatic Society (1, 2, 3); Cane Sprees (1, 2); Numerals (1); Banquet Committee (3); Pep Nite Committee (3); S. E. S. (1, 2, 3); President-elect; Calculus Cremation Committee (2).

Stevens Institute Yearbook, Class of 1939.

[3]Roger Snodgrass, "Historical Society Embraces Reines' Nobel Legacy," *Los Alamos Monitor*, October 29, 2007, http://www.lamonitor.com/content/historical-society-embraces-reines-nobel-legacy.

Fred was also an accomplished gymnast and had participated with classmates in the annual Cane Spree. The campus-wide event included various athletic contests, one of which involved two competitors gripping a single cane. The object was to twist, push, and pull the cane out of the hands of the other. From Fred's college yearbook: "His powerful physique stood him in good stead while he twice became '39's Heavyweight Cane Spree Champion." Decades later, when I came upon the description it stoked an arresting memory.

Sometime after he had become a physicist, while visiting my home, Fred said something about the "center of gravity." It is the point around which the weight of an object is evenly balanced. Scientists also refer to it as the center of mass. He proceeded with a demonstration that perhaps no other nuclear physicist could have equaled. He lay face down on the floor and raised his midsection. He tucked his bent right elbow into his stomach while his palm remained flat against the floor. Gradually he lifted his legs into the air.

Like a magician about to perform an act of wizardry, he then raised his chin and the rest of his upper body from the floor. Fred's prone frame was now 18 inches above the floor. Only the palm of his right hand remained in contact with the surface below. He appeared to be suspended in midair. His arm had become the stem around which his body was balanced. This odd pose seemed surreal. But his judicious positioning had merely exhibited a physical truth. For this posturing to have been possible, his elbow had to have been very near his body's center of gravity.

This pose is known as "the wounded peacock:"

In retrospect I realize that Fred's posturing exemplified a characteristic that he later demonstrated as a classroom teacher. He was a performer whose instructive methods could be totally unexpected and included such methods as whistling, singing, and recitation of poetry.

In 1940, the year before Fred's last at Stevens, he married Sylvia Samuels. Her eclectic interest in literature, art, and cooking complemented his dedication to science.

Sylvia and Fred Reines, Reines Collection.

Fred entered graduate school at New York University in September 1941 and three months later, Japan's December 7th attack on Pearl Harbor drew the United States into World War II. Although Fred was of draft age, his marital and student status provided a deferment from military service. After he received a PhD in physics in 1944, his deferment was continued.

In June, when Fred had largely fulfilled his academic requirements at NYU, my parents and I joined other family members at New York's Grand Central Station. His research apparently had been of interest to administrators of a government program then unknown to the public. We had come to say goodbye to him and Sylvia. They were headed west to a destination that he could not reveal, for work that he could not discuss.

We lingered in the great lobby near the door to the track for their train. I watched intently as the door magically swung open when passengers approached and then closed behind them. My gaze prompted Fred to ask if I knew why this was happening. When I nodded no, he proceeded to unravel the mystery.

He pointed to two tiny windows that faced each other from handrails on the left and right sides of the entryway. One fixture, he explained, emitted a beam of infrared light directed at the other fixture, and although invisible to the human eye, the beam completed an electronic circuit. A traveler approaching the door would unwittingly interrupt the circuit, which activated the door-opening mechanism.

This all seemed strange, since automated supermarket and garage doors were still matters of the future. Ever the science enthusiast, Fred homed in on details. At age 10 this may have been the first time I heard the words electric eye and infrared light. But more than vocabulary, the memory of Fred's exuberant manner of explanation remains vivid.

The following year we learned that Fred's travel destination was Los Alamos, New Mexico, home to the Manhattan Project, which developed the atomic bomb.

After Fred and Sylvia moved, I saw Fred infrequently. Following his years in Los Alamos he held positions at Case Institute in Cleveland and finally at the University of California, Irvine. Through the years we were in touch intermittently, mainly at family celebrations or during one of his occasional visits east. But like other members of our family I followed his career assiduously.

Fred and his young cousin Leonard (me).

Chapter 4

The Einstein Letter, the Hidden City, and the Bomb

On August 6, 1945, the United States dropped a new and devastating bomb over the Japanese city of Hiroshima. Three days later, a similar bomb was released over Nagasaki. The cities were pulverized, and about 200,000 residents were killed by the two bombs, half of them instantly. Germany and Italy, the other two World War II axis powers, had already surrendered to the Allied forces. The fearsome bombs induced Japan to do so as well. With the signing of surrender documents on September 2nd, the war officially ended. Meanwhile, information began to surface about the atomic bomb and the program that led to its development. Much remained secret, but it became clear that Fred Reines was part of that effort. Since his work remained classified, he could say little about it.

The impetus to build the bomb arose from a 1938 experiment in a Berlin laboratory. Otto Hahn and Fritz Strassmann found that bombarding uranium with neutrons could produce the lighter element barium. Hahn, puzzled by the finding, forwarded the information to his former research colleague, Lise Meitner, then in Sweden. Meitner was Jewish and she had recently been forced from her position in Germany. Meitner and her physicist nephew, Otto Frisch — employed in Denmark, but then visiting his aunt — postulated that the uranium nucleus had split into two lighter nuclei. They realized that significant mass would be lost in the process and therefore energy gained, in accordance with Einstein's mass-energy equivalence.

Einstein's famous 1905 equation, $E = mc^2$, held that mass is a form of energy, according to the law of energy conservation. Wolfgang Pauli had used this understanding of energy conservation when he postulated the existence of the neutrino in 1930.

Light c travels at 186,000 miles per second. This means that when even a small amount of mass m is multiplied by c^2, the equivalent energy E can be immense.

In 1933, Leo Szilard had conjectured about the possibility of creating a nuclear chain reaction. When in 1939 he learned about the Hahn–Strassmann experiment, he thought he knew just how a chain reaction might be accomplished. When a uranium nucleus, after bombardment by a neutron, splits apart, it may also release additional neutrons. The freed neutrons shoot about in all directions and some of them will split other uranium nuclei, furthering more neutron bombardments and an ever-growing release of energy. The energy generated from a chain reaction with even a small quantity of uranium could produce a massive explosion.

Szilard worried that after the Berlin experiment, German scientists would try to develop a nuclear weapon. In the summer of 1939, he shared his concern with his friends Eugene Wigner and Edward Teller, who, like Szilard, were originally from Hungary. All were émigré physicists teaching in American universities, and they agreed on the need to alert the U.S. administration. Szilard had a friendship with Albert Einstein, having collaborated with him on research in the 1920s, and he thought that a message from the iconic scientist would be the best approach. The three scientists met with Einstein, who agreed on the need for the U.S. to take action.

Szilard drafted a letter in German, the language best understood by Einstein. Ironically, it was also the language of the anticipated target of the bomb. Wigner translated the draft into English. After further adjustments by Wigner, Szilard, Teller, and Einstein, the final version was completed. Szilard knew Abraham Sachs, an economist who was a friend of President Franklin Roosevelt, and Sachs agreed to convey the message to the president. He was able to meet with Roosevelt in October 1939, weeks after the war had begun with Hitler's invasion of Poland.

Sachs spoke to the president about the immense potential power of an atomic weapon and gave him a copy of the letter signed by Einstein.

Although the U.S. had not yet entered the war, Roosevelt viewed Hitler and the Nazis with contempt. At the message's core was anticipation that the Germans might develop an atomic weapon and that an American atomic weapon would deter the Germans from using theirs.

The letter began by noting that recent work by Enrico Fermi and Leo Szilard indicated that "the element uranium may be turned into a new and important source of energy." As a result, it could "become possible to set up a nuclear chain reaction in a large mass of uranium, by which vast amounts of power and large quantities of new radium-like elements would be generated." This could lead to the construction of "powerful bombs of a new type." The letter cited a need to marshal governmental, private, and university resources to advance an American program.[1]

Albert Einstein and Leo Szilard and the letter they composed with Edward Teller and Eugene Wigner to President Franklin Roosevelt, warning of a potential atomic bomb threat from Germany. Courtesy of the Atomic Heritage Foundation.

The letter concluded with an ominous observation: Germany had stopped the sale of uranium from the Czechoslovakian mines that it had

[1] Richard Rhodes, *The Making of the Atomic Bomb*, NY: Simon and Schuster, 1986: 303–317.

taken over. After hearing Sachs paraphrase the letter, Roosevelt responded: "What you are after is to see the Nazis don't blow us up." Sachs responded, "Precisely."[2] Roosevelt established an advisory committee on uranium, which eventually led to an effort to build an atomic bomb. The effort was named "The Manhattan Project" in 1942 by then-Brigadier General Leslie Groves, its newly appointed director. Stocks of uranium had already been stored at locations in New York City, and Groves' Army Corps of Engineers had an office in lower Manhattan.

Only after the atomic bombs were detonated over Japan did the public learn that three "hidden cities" had become the principal locations for developing the bomb.[3] Two of them, Oak Ridge, Tennessee, and Hanford, Washington, contained secret engineering and chemical plants. They produced the uranium and plutonium that would become the essence of the bombs. The third city, Los Alamos, was a newly built town on an isolated mesa, an hour's drive from Santa Fe, New Mexico. Los Alamos contained the laboratory for the project and became home for those who worked there — scientists, engineers, mathematicians, technicians, administrators, support personnel — and their families.

Secrecy about the project was so pervasive that scientific personnel were told to use pseudonyms when communicating with each other by telephone or in writing. Even the existence of Los Alamos could not be acknowledged. Enrico Fermi (pseudonym, "Henry Farmer"), like many leading scientists, moved there with his family in 1943. After the war his wife, Laura, described the bizarre condition of their residency.

> For two and a half years the city was not marked on the maps, had no official status, was not part of New Mexico, its residents could not vote. It did not exist. That city was Los Alamos to those living there, Site Y to the few outsiders who knew about its existence, and Post Office Box 1663 to correspondents and friends of the inhabitants.[4]

---

[2] "The Einstein-Szilard Letter — 1939," The Atomic Heritage Foundation, https://www.atomicheritage.org/history/einstein-szilard-letter-1939.

[3] Jay Walz, "Atom Bombs Made in 3 Hidden 'Cities'," *The New York Times*, August 7, 1945: 1.

[4] Laura Fermi, *Atoms in the Family*, Chicago: University of Chicago Press, 1954: 206.

The population escalated from a few hundred in 1943 to more than 6,000 two years later. Besides the laboratory, more than one thousand structures had been built, including houses and apartments, administration buildings, military barracks, elementary and high schools, a hospital, recreational facilities, and food and other service facilities.[5]

When Fred arrived in mid-1944 the place had the trappings of a typical small town including a mayor and local governing council.[6] Yet the location was far from other populated areas and was surrounded by a barbed wire fence. Military guards on horseback patrolled areas beyond the town's perimeter. At the main entrance a large sign was posted on the wooden guardhouse: "Passes Must Be Presented to Guards." Inside the secret city another ring of barbed wire encircled the laboratory buildings.

Even if Fred had been informed in advance about the conditions of employment, the junior scientist could hardly have anticipated the dazzling brainpower in this new community. J. Robert Oppenheimer, the director of the laboratory, had recruited scores of talented scientists to Los Alamos. Many in this secret city were Nobel laureates or future laureates.

Oppenheimer had been a distinguished professor of physics at the University of California, Berkeley. Many of the recruited scientists were Jewish refugees or had been other targets of Nazi oppression. Most of them had come to the United States from Europe just before their countries had fallen under Hitler's control. Nobel laureates among them included Niels Bohr, the iconic physicist from Denmark. Bohr had created the model that depicted the atom as a positively charged nucleus orbited by negatively charged electrons. Enrico Fermi from Italy had discovered new elements produced by neutron irradiation and had offered the first theory of beta decay. In 1942 he headed a secret experiment at the University of Chicago that produced the first controlled nuclear chain reaction. An uncontrolled version of this chain reaction would become the explosive core of the atomic bomb. Isidor Rabi, born in Austria though brought up in America, received a Nobel Prize in 1944 while at Los Alamos. His

---

[5] Walz: 1; Thomas Widner, Final Report of the Los Alamos Historical Document Retrieval and Assessment (LAHDRA) Project, Chapter 6, prepared for the CDC, November 2010.

[6] James Gleick, *Genius: The Life and Times of Richard Feynman*. NY: Pantheon Books, 1992: 185.

award was for the discovery of nuclear magnetic resonance, the medical imaging technique used in magnetic resonance imaging (MRI).

Future Nobelists who had escaped from Nazi-controlled Europe included Eugene Wigner from Hungary, Hans Bethe from Germany, and Emilio Segrè from Italy.

Edwin McMillan was among several American future winners. His wife, Elsie, spoke for many when expressing doubt that she would ever again live in a community with "so many brains."[7]

J. Robert Oppenheimer. Courtesy of the Atomic Heritage Foundation.

When Fred arrived, he was awed by the illustrious company. He relished working among what he called "perhaps the greatest collection of scientific talent the world has ever known."[8] His brother David once told

[7] Rhodes, *Making of the Atomic Bomb*: 567.

[8] Reines Nobel Lecture, https://www.nobelprize.org/prizes/physics/1995/reines/biographical/.

me that Fred revered many of his colleagues, but none more than Oppenheimer, for his scientific and managerial brilliance.

The sole purpose of the Manhattan Project was to produce an atomic bomb. The practicality of the goal had been confirmed by Fermi's controlled nuclear chain reaction in 1942. That had been conducted in a makeshift laboratory, a converted squash court beneath the University of Chicago's Stagg Field.

In 1938 Otto Frisch, Lise Meitner's nephew, had named the splitting of the atom "fission," a word that already described bacterial cell division. According to the physics model, the nucleus elongates and divides, as might a drop of liquid. The original spherical drop stretches, becomes thinner in the middle and then separates into two smaller drops.[9]

While at New York University, Fred became interested in this newly identified phenomenon. He ultimately produced a study that expanded on a paper by Niels Bohr and John Wheeler about nuclear fission. His PhD dissertation dealt with the "deformation" of the atomic nucleus and was titled, "Nuclear Fission and the Liquid Drop Model of the Nucleus." The subject was so sensitive that publication was delayed until after the war.

At Los Alamos Fred was assigned to a small group headed by Richard Feynman. Feynman was Fred's age and was little known outside a circle of his peers. Eventually Feynman came to be "regarded as the most brilliant, influential, and iconoclastic [theoretical physicist] in the post-World War II era."[10]

Feynman had been recruited to the Manhattan Project before completing his PhD at Princeton, where his mentor was John Wheeler. He arrived in Los Alamos early in 1943, when the "hidden city" barely had begun to take shape. Laboratory construction was not yet completed. Dormitories were still being built, so people shared rooms in the few

---

[9] "This Month in Physics History," American Physical Society News, December 2007, https://www.aps.org/publications/apsnews/200712/physicshistory.cfm.

[10] Encyclopedia Britannica, https://www.britannica.com/biography/Richard-Feynman.

existing facilities that previously constituted the Los Alamos Ranch School.

By the time Fred joined Feynman's group in 1944 its members were dealing with various challenges related to producing a bomb. The group included Julius Ashkin, Richard Ehrlich, and Theodore Welton. Like Feynman and Reines they were in their 20s, and all would later have successful academic careers.

At Los Alamos they were assessing the variation in explosive yields by different volumes of uranium-235 (U-235). Another task was to evaluate methods of separating U-235 from raw uranium ore. About 99.3 percent of natural uranium is U-238, which does not readily undergo fission. The remaining 0.7 percent is the isotope U-235 which, when bombarded with neutrons, can split into other elements and release energy. (The nucleus of each uranium atom contains 92 protons. The nucleus of a U-235 atom contains 143 neutrons, and of a U-238 atom 146 neutrons.)

Only a bulk of concentrated U-235 could sustain a nuclear chain reaction that yields energy explosively. Current terminology for weapon-quality concentrations of U-235 is "highly enriched" uranium. Extrapolating from the 1938 Berlin lab experiments, the resulting burst of energy could generate a massive explosion. The Manhattan Project included development of both uranium- and plutonium-based bombs.

Details about Fred's scientific contributions to the Manhattan Project are now partially available. Several, though not all, of his reports at the laboratory are unclassified and became "publicly releasable" in 1996. They reveal his interest in the many and varied effects of a nuclear explosion — kinetic energy, heat. fire, radiation. Some of his early reports are filled with terms and mathematics unlikely to be familiar to the general reader. For example, one study conducted by Fred and other group members produced a formula that anticipated effects of the first atomic bomb test, dubbed Trinity, at a prescribed distance. Written by Fred in November 1944, eight months before the test was conducted, it was titled, "Formula for the Critical Radius for One Velocity."

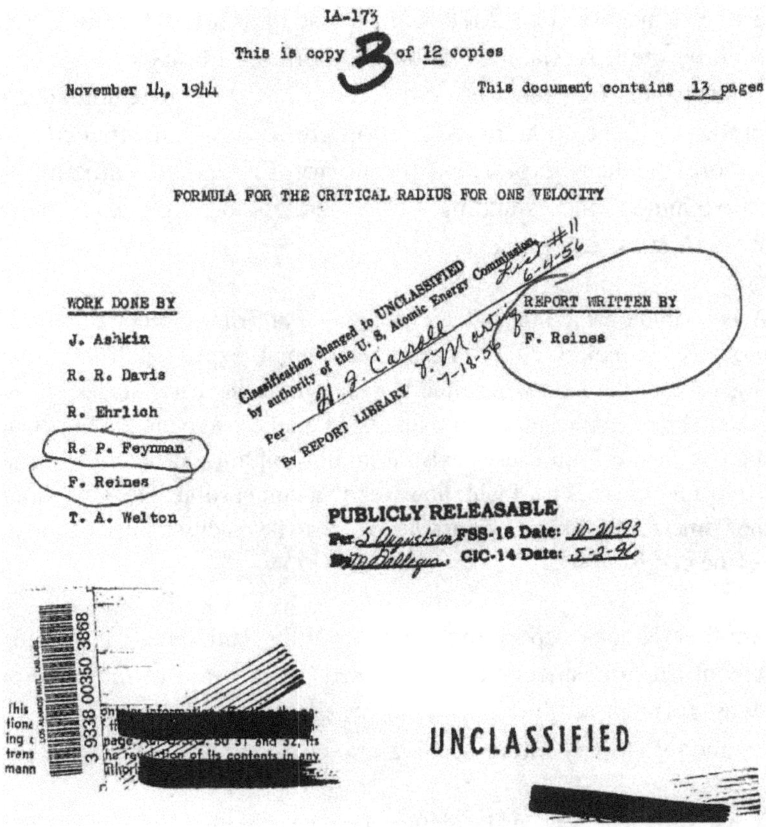

LA-173

This is copy 3 of 12 copies

November 14, 1944 This document contains 13 pages

FORMULA FOR THE CRITICAL RADIUS FOR ONE VELOCITY

WORK DONE BY

J. Ashkin

R. R. Davis

R. Ehrlich

R. P. Feynman

F. Reines

T. A. Welton

REPORT WRITTEN BY

F. Reines

Classification changed to UNCLASSIFIED
by authority of the U. S. Atomic Energy Commission
Per H. J. Carroll 6-4-56
By REPORT LIBRARY 7-18-56

PUBLICLY RELEASABLE
Per S. Augustson FSS-16 Date: 10-20-93
By M. Ballegos, CIC-14 Date: 5-2-96

3 9338 00350 3868

[redacted]

UNCLASSIFIED

[redacted]

Cover of a report by Reines on work by Reines, Feynman, and others, which became "publicly releasable" in 1996. Los Alamos National Laboratory Library.

Another of his reports also includes terminology and equations beyond the scope of the non-specialist. From that report: "Evidence is presented to show that hydride gadgets can be treated with reasonable accuracy in a simple manner despite the complicated nature of the hydrogen scattering." Titled, "Treatment of Anisotropic Scattering by Hydrogen in Multiplying Systems," Fred wrote the report in April 1945. Both studies were conducted in collaboration with Feynman and other members of Feynman's group. These and other declassified reports by

Reines were accessed in 2018 from a list of Reines' technical reports provided by the Los Alamos National Laboratory library.

Other papers by Fred were more comprehensible to a non-specialist. In October 1945, he co-authored a report on work he had conducted titled, "July 16th Nuclear Explosion: Incendiary Effects of Radiation." His report expanded understanding of the effects of high level radiation exposure. A summary passage:

> A few simple experiments were performed at Trinity to determine the probable incendiary effect of the atomic bomb explosion. The results suggest that the risk of fire due to radiation in the early stages of the atomic bomb explosion is very small and is likely to be much less than the risk of fire from causes existing in the buildings at the time of the explosion. There is no doubt, however, that combustible material within the flame zone from the explosion is likely to be readily ignited and may set the debris on fire.

Another Reines report that month, "July 16th Nuclear Explosion: Permanent Earth Displacement," assessed the creation of the large crater that was formed at the Trinity blast site. The findings were easily understandable to any interested reader:

> A measurement was made of the permanent earth movement in the neighborhood of the tower caused by the nuclear explosion at Trinity. The crater was apparently formed by a compression phenomenon [which] we noted to be much shallower than craters formed by ordinary explosives. Scaling up existing information on the radius of the craters, the TNT equivalence of the nuclear bomb is given as 10,000 tons to within 50 percent. Observations were made on the damage in the crater region and suggestions are given as to the nature of structures which might be expected to withstand the blast close to the gadget [the nuclear device]. It is also noted that, despite previous opinion to the contrary, it is apparently possible to reduce greatly the volume of earth blown away by the blast by suitably protecting the ground.

Since the reports remained secret until 1996 they were never discussed publicly by Reines or Feynman. Fred died in 1998 and Feynman in 1988.

But the dynamism that Feynman brought to discussions with Fred and other group members may be gleaned from a 1976 article by Feynman about his Los Alamos experience.[11]

Richard Feynman

Frederick Reines

Security Badge Photographs c. 1944, Los Alamos National Laboratory. Compiled by Alan B. Carr, Laboratory Historian.

He described his meeting with Niels Bohr and Bohr's son Aage, whose code names were Nicholas Baker and James Baker:

> We were at a meeting once, the first time he came, and everybody wanted to see the great Bohr. So there were a lot of people there, and we were discussing the problems of the bomb. I was back in a corner somewhere. ...
>
> In the morning of the day he's due to come next time, I get a telephone call. "Hello Feynman?" "This is Jim Baker." "It's his son."
>
> "My father and I would like to speak to you." "Me? I'm Feynman, I'm just a —"
>
> "That's right. OK." ....

[11] Richard P. Feynman, "Los Alamos from Below: Reminiscences of 1943–1945," *Engineering and Science*, January–February, 1976: 28, http://www.f.waseda.jp/sidoli/Feynman_2003.pdf.

> So [we] go into an office in the technical area and he says, "We have been thinking how we could make the bomb more efficient and we think of the following idea." I say, "No, it's not going to work. It's not efficient. Blah, blah, blah." ....
>
> Then the son told me what happened. The last time he was there, he said to his son, "Remember the name of that little fellow in the back over there? He's the only guy who's not afraid of me, and will say when I've got a crazy idea. So next time when we want to discuss ideas, we're not going to be able to do it with these guys who say everything is yes, yes, Dr. Bohr. Get that guy and we'll talk with him first."

Feynman's unvarnished vernacular and unrepressed brilliance must have impressed Fred as it had Bohr. Being in Feynman's small group meant frequent exposure to the swirl of his ideas, his boldness, and his uninhibited manner of expression.

Fred admitted to being dazzled by the brilliance of many of the Manhattan Project scientists. But he never publicly discussed the Feynman group activities, which remained classified. Still, during his first year in Los Alamos, Fred worked closely with Richard Feynman and it would be surprising if he were not influenced by Feynman and his sense of derring-do. Indeed, years afterwards Fred cited derring-do as a most admirable trait of a scientist.

Feynman later became world-famous. He was awarded the Nobel Prize in 1965 for his work on quantum electrodynamics. But when he was in Los Alamos, as he later said, "I was not anybody famous at all. ... I worried about no big decisions. I was always flittering about underneath."[12]

Feynman's group was part of the overall theoretical division. Composed of about 100 scientists, the division was headed by Hans Bethe. Feynman likely considered Bethe a member of the higher echelon that made important decisions. In addition to U-235, plutonium-239 was seen as a possible core for a fissile bomb. As the head of T Division, Bethe promoted the development of a method to detonate a plutonium-based bomb. Described as the implosion method, it initiated a chain reaction that would last about a microsecond.[13]

[12] Ibid.: 12.

[13] Hans Bethe, The Implosion Design of the Plutonium Atomic Bomb, https://www.webofstories.com/play/hans.bethe/93;jsessionid=212F85AE33BED01337D0448B2D6CB02B.

Bethe had also taken an interest in the presumptive neutrino long before Fred focused on it as a subject of research. Prior to the Manhattan Project, Bethe had produced several papers related to the neutrino. After Fermi's 1934 description of beta decay and inverse beta decay, Bethe and Rudolph Peierls published a letter in *Nature*. The two physicists calculated the neutrino cross-section. They concluded that the probability of a neutrino interacting with matter was fantastically remote: less than one chance in several billion. "This meant," they wrote, "that one obviously would never be able to see a neutrino."[14]

Hans Bethe

Rudolph Peierls

Courtesy of the Atomic Heritage Foundation.

After Reines and Cowan's discovery, Fred reminded Bethe of his and Peierls' 1934 pronouncement. "Well," Bethe replied whimsically, "you shouldn't believe everything you read in the papers." Peierls separately acknowledged that he and Bethe had faltered. They had not anticipated "the existence of reactors producing neutrinos in vast quantities," he said. In an implicit bow to Fred and others, he added that neither had they sufficiently allowed for "the ingenuity of experimentalists."[15]

In 1944, it is unlikely that Bethe, Reines, or anyone else at the Los Alamos laboratory was thinking much about the neutrino. They were

[14] HA Bethe, RE Peierls, "The Neutrino," *Nature*, April 7, 1934: 532.

[15] Jayawardhana: 72; Sutton: 34–35.

focused on the all-consuming task of building an atomic bomb. Still, it was understood that the explosion of an atomic bomb would produce a huge flux of neutrinos.

After the war ended, most of the Manhattan Project scientists left Los Alamos. Many returned to earlier posts or found new ones in universities or industry. Fred opted to remain at the laboratory. In the brief time he had been there he had gained respect as a thoughtful and productive investigator.

Frederick Reines, c. 1945. Courtesy of Rebecca Collinsworth, Archivist, Los Alamos History Museum.

In January 1947, he wrote about the presumed radioactive consequences of a test explosion atop differently sized towers, "Spread of Radioactive Contamination from a Tower-Burst Atomic Bomb." The paper noted that fission products included "prompt and delayed" emission of gamma rays and neutrons. Neutrinos were not mentioned in the paper, though a nuclear explosion was also assumed to emit a large flux of the particles. Years later Fred wrote: "The idea of searching for the elusive neutrino had, in fact, occurred to me as early as 1947, but the opportunity did not present itself."[16] Another four years would pass before he and Clyde Cowan would begin their search.

[16] Reines Nobel Lecture.

Chapter 5

From Bombs to Neutrinos

The two bombings in Japan had been preceded by the secret July 16th atomic test, code-named Trinity. All three detonations were based on fissile reactions and each yielded an explosive equivalent of 15–20 kilotons (thousand tons) of TNT. The Trinity and Nagasaki explosive cores were plutonium and the Hiroshima core uranium.

Trinity was conducted 70 miles north of Alamogordo, New Mexico and 120 miles south of Albuquerque, in the 275-square mile White Sands desert. Earth-covered bunkers containing monitoring equipment had been built at varied distances from ground zero. About 250 scientists and other observers were stationed at distances as far as 20 miles from the bombsite and they had been given dark welder's glasses to wear during the test. Anxiety was rife. Some preliminary limited tests had failed and despite the promising calculations for Trinity, it too could end as a spectacular dud. Conversely, a few scientists reportedly wondered whether the nuclear chain reaction that set off the explosion could also set the air on fire. Whether anyone actually thought that was a real possibility, is doubtful. But the extent of damage from an explosion was uncertain and some did worry about their safety.

Elsie McMillan knew that her husband Edwin and the other scientists would be testing the atomic bomb. She later recounted his words to her before his trip to the test site:

> We ourselves are not absolutely certain what will happen. In spite of calculations, we are going into the unknown. [We could] all be blown to bits, if it is more powerful than we expect.

She recalled the moment when Edwin returned home safely after the test:

> The door opened about 6:00 o'clock in the evening. We were in each other's arms. Then, and only then, did the tears come streaming down my face.[1]

For those unable to hear the radioed countdown, a warning rocket was fired one minute in advance of the scheduled detonation. A loud gong was sounded at the 10-second mark. The countdown came to an end. The pre-dawn sky burst into blinding light. The flash could be seen two hundred miles away. The sound that reached observers erupted as if a hundred locomotives were roaring in unison. It followed a shock wave of hurricane force and a rush of engulfing heat.

Descriptions by scientists who witnessed the Trinity test were steeped in emotion:

> Kenneth Greisen: "I was personally nervous, for my group had prepared and installed the detonators, and if the shot turned out a dud, it might possibly be our fault."
>
> Joan Hinton: "It was like being at the bottom of an ocean of light. We were bathed in it from all directions."
>
> Edwin McMillan: "The whole spectacle was so tremendous and one might almost say fantastic that the immediate reaction of the watchers was one of awe rather than excitement."
>
> Robert Serber: "The grandeur and magnitude of the phenomena were completely breath-taking."[2]

After the war, Fred Reines wrote a lengthy letter to the Stevens alumni magazine about his experience.[3] He hoped that the bomb would "shred ... complacency about the necessity for peace on earth." He commented on

---

[1] Elsie McMillan, Voices of the Manhattan Project, https://www.manhattanprojectvoices.org/oral-histories/elsie-mcmillans-lecture.

[2] Atomic Heritage Foundation, https://www.atomicheritage.org/key-documents/trinity-test-eyewitnesses.

[3] Frederick Reines, "Atomic Bomb Experiment," *Stevens Institute Alumni Magazine*, October 1945: 15–16.

the process of nuclear fission and the challenge of separating the U-235 isotope from the more plentiful U-238. As for the mechanism to initiate a nuclear explosion, "I am compelled to keep silent on this point for reasons of security." Then he described what he and the other Los Alamos scientists had witnessed.

> This was absolutely the most awe-inspiring spectacle man has yet created on this earth. Even though we who worked on the test preparations knew what to expect in the way of a spectacle, we were left gasping by the magnificent sight that spread out before us.

Prior to the test Fred had spent days installing apparatus to monitor the bomb's impact at various distances from ground zero. The detonation was scheduled for 4 a.m. on July 16. Hours earlier, lightning storms in the area had raised the possibility of postponement. As last-minute preparations continued, Fred became anxious. He worried "that a lightning strike on the tower might set off the bomb and blast us into oblivion." At 2 a.m. he and others moved to a base camp ten miles from the bomb. They learned that the test would still go forward, though later than planned. At 5:20, just before dawn, a voice on the public-address system announced that the firing would occur in 10 minutes.

> [We were instructed] to lie on the ground, facing away from the bomb. The instructions continued, not to look at the bomb even through our welder's glass.... The unbearable agony of waiting mounted and people started muttering nervous remarks.... 5 minutes. 4 minutes. The countdown continued. 1 minute, then 45 seconds.

As the countdown reached zero a blinding flash bleached the entire panorama. The light intensity, he wrote, was "many times that of the noon-day sun." When the intensity began to diminish Fred rolled over and looked at the explosion through his dark welder's glass. He saw a bright ball moving upward. He removed his glass and could see that the ball was "a beautiful purple" that began changing into a spectrum of colors. "It rose to about 40,000 feet and mushroomed out forming a huge grey cloud." Then came the blast wave, a rumbling sound and a feeling of pressure on his ears.

A short period of awed silence followed and then suddenly everyone started cheering, shaking hands, and exchanging comments of amazement at the "miracle" they had just witnessed.

Trinity Atomic Bomb Test, July 16, 1945.

Note: This image is a work of a U.S. Department of Energy (or predecessor organization).... As a work of the U.S. federal government, the image is in the public domain.

Fred suggested that Stevens graduates could help develop practical uses for atomic power. He ended his letter with the hope that students could now continue their education in a peaceful world.

Months later, when visiting back east, he told me that he had been especially surprised by an effect on the ground. Much of the desert surface near the detonation site had turned into green glass. The intense heat had fused surrounding particles of sand. The newly formed glass was mildly radioactive and given the name trinitite. The presence of radioactivity limited the time that the scientists could spend in the area and for decades, elevated levels of radioactivity precluded public visits to the site.

Twenty years after the Trinity detonation, Oppenheimer said that among its observers, "A few people laughed, a few people cried, most people were silent."[4] His description suggested far less enthusiasm about the event than reported by Fred. For Oppenheimer silent awe was the

[4] James Temperton, "Now I Become Death ... Oppenheimer's Infamous Quote," *Wired*, August 9, 2017, https://www.wired.co.uk/article/manhattan-project-robert-oppenheimer.

predominant recollection, not the cheers and handshakes that followed. But other observers aligned with Fred. Ralph Carlisle Smith, who later became assistant director of the Los Alamos laboratory, wrote that after the detonation "there was a spontaneous cheer from the observers." Kenneth Greisen said that "between the appearance of light and the arrival of the sound, there was loud cheering [and] we all went about congratulating each other and shaking hands."[5]

Perhaps Oppenheimer's latter-day emphasis reflected his own attitude, one of greater worry than celebration. He said later that while viewing the explosion he thought of a passage from the Hindu scripture, the Bhagavad-Gita. The poetic passage suggests the potential for sheer horror to which humankind had just become privy: "Now, I am become Death, the destroyer of worlds." Some have interpreted his reference as an expression of immediate regret for building the bomb.[6] At the time of the test, Oppenheimer had not voiced the Bhagavad-Gita passage. His brother Frank, who was next to him in their bunker, recalled Robert saying only, "It worked."[7] Nor did Oppenheimer oppose the decision to drop the bombs on Japan.

In June 1945, prior to the Trinity test, James Franck, a Nobel physics laureate, presented a secret report to President Truman's Interim Committee, an advisory committee on nuclear affairs. Signed by seven prominent nuclear scientists, the Franck report recommended against a surprise atomic attack against Japan. Rather, the United States should invite Japanese officials to view a bomb detonation in a barren location. Fear of being a target would presumably induce Japan to end the war and would discourage others from attacking the United States. The report's authors also hoped that a demonstration would convince nations to establish international control of such weapons.[8]

The Interim Committee rejected the proposal for a demonstration, and the rejection was supported by a panel of leading scientists including

---

[5] Ralph Carlisle Smith, https://www.atomicheritage.org/profile/ralph-carlisle-smith; Trinity Test-1945, https://www.atomicheritage.org/history/trinity-test-1945.

[6] "Did Oppenheimer Regret His Invention?" *Quora*, https://www.quora.com/Did-Oppenheimer-regret-his-invention-What-happened-after-the-bombing-that-contributed-to-his-downfall.

[7] Alex Wellerstein, Nuclear Secrecy Blog, http://blog.nuclearsecrecy.com/2014/05/23/oppenheimer-gita/.

Oppenheimer and Fermi, who wrote, "We see no acceptable alternative to direct military use."

Leo Szilard had signed the Franck report. In a 1986 oral history, Edward Teller recalled that Szilard asked him to solicit signatories urging a demonstration bomb. Teller was inclined to do so, he said, but first wanted to check with Oppenheimer.

> Oppenheimer talked me out of it, saying that we as scientists have no business to meddle in political pressure of that kind. We know too little about that. I am ashamed to say that he managed to talk me out of the right intention. I then wrote Szilard a letter saying that in a way I agreed with him, but I feel I should not act. And that I regret.[8]

After the Trinity test Szilard produced a petition that he hoped would be presented to President Truman. It urged the United States to offer surrender terms to Japan that would assure its citizens "a life devoted to peaceful pursuits." Only if Japan then refused to surrender should the bomb be considered. Most of the 70 co-signers were junior personnel; Oppenheimer's name was not among them. The petition never reached the president.[9]

On August 14th, one week after Nagasaki was bombed, Japanese radio announced the emperor's acceptance of the Allies' terms of unconditional surrender. World War II had effectively ended. The surrender documents were signed on September 2nd aboard the USS Missouri in Tokyo Bay.

Americans celebrated. Qualms about death and destruction wrought by the new weapons were quelled by joyous relief that the war was over. By some estimates, an invasion of Japan would have cost a half-million lives or more.

Several who had helped build the bomb felt uneasy about their participation in a project that could cause such destruction. But the pervasive attitude among the participating scientists, including Fred, was shared by the general public when it learned about the bomb. Americans

---

[8] "Edward Teller's Interview," Voices of the Manhattan Project, https://www.manhattanprojectvoices.org/oral-histories/edward-tellers-interview.

[9] Szilard Petition, http://www.dannen.com/decision/45-07-17.html.

and their allies were thankful that the new weapon had abruptly brought the war to an end.

In the 1960s some historians began to argue that dropping the bombs had been unwise. They contended that projected casualty rates from an invasion had been exaggerated and some believed that the Japanese were ready to surrender in any case. Seventy years after the war the issue was still being debated.[10]

Work on atomic weaponry continued in the post-war period. Efforts were ramped up with the onset of the Cold War and with the Soviet Union's detonation of an atomic bomb in 1949. (The Soviet success was abetted by spies for the Soviet Union. Prominent among them was Klaus Fuchs, a scientist who had worked in Los Alamos on the Manhattan Project.)

Tensions had also surfaced about whether the United States should seek to construct a thermonuclear weapon. Based on the fusion of hydrogen nuclei, a hydrogen bomb would be far more powerful than the fission-based device. Edward Teller supported, and Oppenheimer opposed, U.S. development of the "H-bomb." The issue reached a climax when the government's Personnel Security Board held secret hearings in 1954 about whether Oppenheimer's security clearance should be renewed. Gordon Gray, the board's chairman, presided. Teller testified that he did not question Oppenheimer's loyalty but nevertheless felt uncertain about his reliability. The caveat was viewed as important to the board's decision not to renew Oppenheimer's clearance.

The proceedings of the hearings became public the next year and many in the physics community distanced themselves from Teller. Fred revered Oppenheimer and also respected Teller. His consternation about the issue was clear in a letter to Teller, dated March 5, 1955.

> Dear Edward,
>
> I read your story of the Super development some days ago but withheld comment until now in order to think about it....

---

[10] Debate Over the Japanese Surrender, https://www.atomicheritage.org/history/debate-over-japanese-surrender.

It is good that you wrote it. However, inextricably intertwined with the H-bomb story is the revocation of Oppenheimer's clearance and your testimony before the Gray board. One can hope that the heat of your struggle with varying points of view on the H-bomb has abated sufficiently so that you can see the basic confusion as to the meaning of security clearance in your argument against his clearance....

A gesture which will heal the festering sore that is the Oppenheimer case and would in my view demonstrate a greatness of spirit in keeping with your masterful contributions to the thermonuclear field would be to work for the reinstatement of his clearance. I realize full well that such efforts might not be compatible with your conscience but I can only suggest it in the hope that it is.

Most sincerely,
Fred

Evidently Teller did not take Fred's advice, nor is there evidence that he even responded to the letter.

Edward Teller. Courtesy of the Atomic Heritage Foundation.

Some alumni of the Manhattan Project refused to engage in further weapons research as a matter of principle. Most simply left Los Alamos in pursuit of careers elsewhere. But several of them continued to consult on the weapons program, understanding that the Soviet threat justified their efforts.

Fred was among the reduced number who remained fulltime at the Los Alamos Laboratory, and he soon became a group leader in the theoretical division. In March 1947, Fred was introduced to Harris Mayer, who had just arrived at the laboratory.

Mayer grew up in New York City. After receiving a bachelor's degree from New York University in 1940, he enrolled in the graduate physics program at Columbia University. There, in 1942, he joined in work related to the Manhattan Project under the direction of future Nobel laureate Marie Goeppert-Mayer (no relation) and later, Edward Teller. The two renowned physicists moved to the University of Chicago after the war and Harris went with them. Under Teller's supervision at Chicago, Harris received a PhD in 1946.

Some 70 years later, Mayer reflected on his subsequent friendship with Fred. At age 97, with an impressive memory for details, he recounted their long history of shared experiences. The Mayer and Reines families had become close friends.

> My wife Rosalie and I, and Sylvia and Fred, almost immediately after we met, found we had liked each other. We were in Los Alamos where there were no parents and no cousins. We were only with ourselves. So we became like our own family.[11]

The Mayers, who had not yet had children, lived in an apartment near Fred and Sylvia on Trinity Place. In 1948, when Sylvia was about to give birth to their second child, Fred wanted to stay with her in the hospital. He asked Harris and Rosalie for a favor. Would they please move into the Reines' apartment for a few days and take care of 3-year-old Robert? "We agreed, happily," Harris said. The experience further bonded the two families, who remained friends long afterwards.

Being with Fred was a joy, Harris recalled, whether socially or professionally. The two men worked together on numerous laboratory

---

[11] Harris Mayer interviews, 2017–2018.

projects. "Fred had a lot of ideas and I had a lot of knowledge about how to develop them. It was sometimes like two brains working as one mind." Harris extolled Fred's ability to take charge and drive toward a goal. He was a natural leader, Harris said, recalling that even as a youngster Fred was a leader, an Eagle Scout.

Mayer said that when he arrived at Los Alamos, the work there seemed unfocused. The atomic bomb had been built and used, but no one seemed to know what the laboratory should do next.

> And here's where Fred's humanity and leadership became evident. He realized that all we knew was how to make a bomb.... It could explode. What he said was, we must understand what it does when it exploded. The military had no idea about this. They didn't know the physics. What Fred did, almost by himself, was [seek to understand] what it does to "nature." And he found out exactly what happened and how it would operate and the damage it would do. He and I worked together on this.

They studied the effects of nuclear bombs with various explosive yields at different distances from points of detonation. Mayer demurred from describing details because, he believed, some of the information was still classified. Fred's initiatives in part paralleled studies underway elsewhere about the health effects of the intense heat and radiation exposure.

Even while developing information of military interest, Fred was devoted to finding ways that nuclear power could be harnessed for peaceful purposes. He wrote an article in 1950 titled, "Are There Peaceful Engineering Uses of Atomic Explosives?" It appeared in the *Bulletin of the Atomic Scientists*. That edition carried not only his article, but others by luminaries including Winston Churchill and Frederic Joliot-Curie. Fred did not advocate the use of atomic explosions to move large masses of earth because of the radioactive aftermath. But he allowed that an explosion underground in certain rock formations could confine the radioactivity, which eventually would dissipate. Still, he wrote, the uses of atomic explosives for large engineering projects at this time "appear at best to be very limited in scope."[12]

---

[12] Frederick Reines, "Are There Peaceful Engineering Uses of Atomic Explosives?" *Bulletin of the Atomic Scientists*, June 1950.

Three years later, in 1953, President Dwight Eisenhower proposed an initiative called "Atoms for Peace." He urged the creation of an International Atomic Energy Agency (IAEA) that would devise methods —

> to serve the peaceful pursuits of mankind. Experts would be mobilized to apply atomic energy to the needs of agriculture, medicine and other peaceful activities. A special purpose would be to provide abundant electrical energy in the power-starved areas of the world.[13]

The IAEA was created in 1957. Fred supported the goals of the agency and in 1958 was a delegate to the Atoms for Peace Conference in Geneva. The agency continues to exist but principally as a monitor of nuclear activities around the world. The lofty peaceful ambitions pronounced by Eisenhower received far less attention.

American open air nuclear testing continued after the war, mainly in the Southwest United States at a desert site in Nevada. Tests were also conducted on atolls in the Marshall Islands, 2600 miles southwest of Hawaii. In 1951, Fred led Operation Greenhouse on Eniwetok atoll. The project included setting off a series of limited fission explosions (uranium- and plutonium-based) as potential triggers for a fusion reaction — a hydrogen bomb.[14] The explosive force based on fusion of the nuclei of hydrogen atoms was expected to be hundreds of times greater than that of the Hiroshima or Nagasaki fissile bombs. After the Greenhouse tests ended in May, Fred returned to Los Alamos. The following year, in November 1952, the first hydrogen bomb was detonated at Eniwetok. The anticipated result was realized. The bomb produced a yield equivalent to 10 megatons (million tons) of TNT.

Contributing to the theoretical underpinnings of Greenhouse was a young physicist, Kenneth Ford, who had arrived in Los Alamos in 1950. His PhD supervisor, Princeton Professor John Wheeler, had suggested that Ford suspend his studies for an interlude at the Los Alamos laboratory. Ford served as a research scientist there for a year before returning to Princeton. While in Los Alamos, Ford met Fred. Although Ford never

[13] Dwight D. Eisenhower, Atoms for Peace Speech, International Atomic Energy Agency, 1953, https://www.iaea.org/about/overview/history.

[14] William Kropp, Jonas Schultz, and Henry Sobel, *Frederick Reines, 1918–1998, A Biographical Memoir*, Washington, DC: National Academy of Sciences, 2009: 10–12.

traveled to the Pacific, his calculations were part of the Greenhouse program.

As Ford recalls, Fred had moved from his previous position in the theoretical division at the Los Alamos laboratory to the test division. Fred's responsibilities in Operation Greenhouse included oversight of diagnostic experiments for nuclear tests. "This was challenging work," Ford said, "since measurements had to be recorded in milliseconds or microseconds before everything, including the measurement apparatus, got blown up."[15]

When I met Ford in 2017, he was very pleased to speak about Fred. He remembered him fondly and described his enthusiasm as infectious. Ford, active and agile at 91, was among few living peers from Fred's time at Los Alamos. When we met, he was still maintaining an autobiographical website and had recently published his 10th book, *Building the H Bomb: A Personal History*.

After Operation Greenhouse ended in 1951, Fred began a sabbatical while remaining in residence at Los Alamos. At the same time, Ford returned to Princeton to complete his PhD and to work on Project Matterhorn, a program there to develop controlled fusion energy. The casual association between Fred Reines and Ken Ford ended for the time being. Later it would materialize into a closer friendship that would prompt a dramatic change in Fred's career.

Meanwhile, Fred began his leave from weapons-related assignments. For months he contemplated nonmilitary physics projects to pursue in the coming years. It was during this time that he decided to attempt the observation of the neutrino.[16]

Fred's years of graduate study, followed by work on nuclear weapons, provided him with a relevant base of knowledge. His interest in neutrino observation was framed by an understanding that a detonated atomic weapon could be a ready source of the particles.

In a chance conversation in 1951 with Clyde Cowan, a Los Alamos colleague, Fred discussed the possibility of tackling a "real challenging problem." Cowan was interested. Fred suggested that they work on the

---

[15] Kenneth (Ken) Ford interviews, 2017–2019.

[16] Reines Nobel lecture.

neutrino. "Great idea," Cowan responded. "He knew as little about the neutrino as I did but he was a good experimentalist with a sense of derring-do," Fred thought. "So we shook hands and got off to working on neutrinos."[17]

The next few years were devoted to a search for the elusive particle. In 1953, they found suggestive, though not definitive, evidence of detection. After further experiments, in 1956 Reines and Cowan published their findings, which confirmed their detection of the neutrino. In 1957, Cowan moved to Washington, DC to join the faculty at George Washington University. A year later he became a professor of physics at nearby Catholic University.

Fred left Los Alamos in 1959 to become chairman of the department of physics at the Case Institute of Technology — later Case Western Reserve University — in Cleveland. He remained there for nearly seven years before moving to the University of California, Irvine.

The story of the neutrino is one of scarcely imaginable extremes. The particle is omnipresent yet proof of its existence had long eluded confirmation. Other than photons — particles of light — neutrinos are the most abundant particles in the universe. The tiniest of entities with mass, they are so plentiful and so penetrating, that every second, billions pass through every cubic inch of every human being. They pass not only through people, but through objects of any kind including the planet Earth, on their way to the farthest reaches of the universe.

Why should anyone care about the neutrino? The answer is threefold. The first is personal. Even if unfelt, knowing that trillions of particles are continuously moving through one's being should intrinsically interest us. Included among those trillions at any moment, are thousands of primordial neutrinos. They were created 13.8 billion years ago, one second after the universe was born. A moment's reflection inspires a sense of awe about such a wondrous phenomenon.[18]

---

[17] *Ibid.*

[18] Shannon Hall, "Astronomers Indirectly Spot Neutrinos Released Just 1 Second after the Birth of the Universe," *Scientific American*, December 1, 2015, https://www.scientificamerican.com/article/astronomers-indirectly-spot-neutrinos-released-just-1-second-after-the-birth-of-the-universe/.

A second reason to care arises from intellectual curiosity in general. Filling gaps of knowledge about natural phenomena can be psychologically satisfying. Every child's learning experience includes asking Why? The question is also a driving force in science. The answer to Why? in the context of basic science is often no more complicated than: To Know. When Henry David Thoreau was asked why he persistently sought knowledge, his reply was a question: "What else is there in life?"[19]

Finally, caring may also have palpable effects. The quest To Know can be a portal to tangible rewards. For a researcher this could mean fame, riches, even a Nobel Prize. Beyond gratification for an investigator, discovery may lead to expansive benefits for humankind: cures for disease, improved transportation, enhanced energy production. The 2015 Nobelist Arthur McDonald was asked what practical benefit might come from his proof that neutrinos oscillated and therefore had mass. He responded without hesitation. Neutrinos, he said, are part of the process that generates enormous energy in the core of the sun, and this energy, from the fusion of hydrogen nuclei, affects our solar system and beyond. Fully understanding the role of neutrinos in the process could advance efforts to develop fusion-driven power on earth.[20]

Even if no practical benefit from a finding is initially obvious, history is replete with unanticipated outcomes. None is more stunning than the relationship of mass and energy formulated by Albert Einstein in 1905. His equation, $E = mc^2$, is based on an understanding that matter and energy may be converted from one form to the other. For more than three decades scientists largely accepted the truth of the equation though not necessarily that the knowledge could be applied in any useful manner. Among the doubters was the great physicist himself. In 1934, Einstein declared: "There is not the slightest indication that [nuclear energy] will ever be obtainable. It would mean that the atom would have to be shattered at will."[21]

---

[19] F.B. Sanborn, *The Life of Henry David Thoreau,* NY: Houghton Mifflin, 1917.

[20] "Physics Nobel Laureate Arthur McDonald Hopes Work Could Pave Way to Fusion," *The Straits Times*, October 7, 2015, https://www.straitstimes.com/world/americas/physics-nobel-laureate-arthur-mcdonaldhopes-work-could-pave-way-to-fusion-power.

[21] Quoted in "Atom Energy Hope Is Spiked by Einstein / Efforts at Loosing Vast Force is Called Fruitless," *Pittsburgh Post-Gazette*, December 1934: 1.

A mere 4 years later, in 1938, the presumption was up-ended with the splitting of the atom in the Hahn–Strassmann laboratory. Another example, of course, was Pauli's initial belief that his postulated neutrino could never be detected. Nearly 60 years later Fred reviewed the pathway of his and Cowan's ultimate success. It also revealed a stubborn characteristic of Fred's persona that fueled his determination. He acknowledged as much:

> So why did we want to detect the free neutrino? Because everybody said, you couldn't do it. Not very sensible, but we were attracted by the challenge.[22]

---

[22] Reines Nobel Lecture.

Part III

Discovery

Chapter 6

Explosive Ideas

The atomic age introduced not only a mammoth destructive force, but also hope that the force would become a beneficent source of energy. Initial post-war assessments of harnessed atomic energy were rapturous. *New York Times* reporter William Laurence declared that with the new energy, "We can turn deserts into blooming gardens. We can air condition the jungles and make the arctic wastes livable … and we can lick disease." Others suggested that an automobile, a train, even a large ocean liner might be fueled by a nuclear pellet the size of a pea.[1]

These descriptions proved to be more fantasy than possibility. Meanwhile, in addition to bomb production, the Los Alamos laboratory had become focused on understanding more fully the effects of an atomic detonation.

In addition to its kinetic effects, a nuclear device could cause injury by fire, intense heat, and high levels of radiation. Understanding the full consequences of atomic weaponry had only begun. Post-war studies sought information about the effect on large human populations as well as physical structures, military equipment, and the positioning of battlefield personnel.

In a retrospective presentation about early attitudes at the Los Alamos Laboratory, Fred said they were permeated by a "can-do spirit."

[1] Spencer R. Weart, "The Heyday of Myth and Cliché," *Bulletin of the Atomic Scientists*, August 1985: 38–43.

> Whenever we thought about new projects, the idea was to set the most interesting (and fundamental) goal without initial concern as to feasibility or practical uses. We could count on the latest technology being available to us at Los Alamos as a result of the instrumentation needs of the weapons program, and that fed our confidence. To his credit, Norris Bradbury, the Director who took over after Oppenheimer, lent enormous support to surrounding the nuclear weapons effort at Los Alamos with a broad scientific and technological base.[2]

Besides designing and assessing weapons, Fred and other Los Alamos scientists retained a core interest in fundamental physics. They frequently discussed the feasibility of experiments devoid of military purpose.

How does one devise an experiment to detect a neutrino, a particle that may not even exist? Or, if it does exist, is invisible, carries no electric charge, and possibly lacks mass? In fact, seeing or capturing an item is not always necessary to prove its existence. As physicist Richard Garwin has noted, gravity and radiation are not visible or subject to capture, yet we know they exist by way of their effects on certain objects.[3]

In the summer of 1951, following his return to Los Alamos from Eniwetok atoll, Fred took a sabbatical from his laboratory responsibilities. After years of working on nuclear bombs, he wanted to pursue a project involving fundamental physics. But he was frustrated. Months into his sabbatical he was still "searching for a meaningful question worthy of a life's work."

Amid his frustrations, he pondered an idea based on his knowledge of fission bombs. An atomic explosion was believed to produce a large flux, or quantity, of neutrinos. Could he devise a means of detecting the particle that many scientists believed was undetectable? Might this be the route to fulfilling his ambition since high school, to be an extraordinary physicist? Or his aspiration, expressed during college, to make a career-building discovery?

---

[2] E.C. Anderson, "The Reines-Cowan Experiments, 1953–1956: Detecting the Poltergeist," *Los Alamos Science*, No. 25, 1997, http://permalink.lanl.gov/object/tr?what=info:lanl-repo/lareport/LA-UR-97-2534-02.
[3] Richard Garwin interview, 2017.

In 1934, Enrico Fermi formulated the theory of beta decay. It held that a neutron could change into a proton, electron, and the small particle of energy posited 4 years earlier by Wolfgang Pauli. Fermi's theory held that the reaction could also work in reverse. Inverse beta decay meant that the particle could combine with a proton and electron to form a neutron and a positron (positive electron). He called the particle, as suggested by his colleague Amaldi, a "neutrino" — little neutral one. (The particle produced by beta decay had been shown to be an antineutrino. Its spin is opposite that of a neutrino, but it is otherwise qualitatively identical.)

In 1938, Fermi was awarded the Nobel Prize in physics for his work on nuclear reactions. By then, under the influence of Nazi Germany, Italy's fascist government had begun to impose anti-Semitic laws. Fermi's wife, Laura, was Jewish, and they fled after the Nobel ceremony from Sweden to the United States. Fermi's brilliance was appreciated in America. In December 1942, he led a team that produced the first controlled nuclear chain reaction, at the University of Chicago. He then moved to Los Alamos where he remained a key figure in the Manhattan Project. After the war he returned to Chicago, but he continued to spend summers at the Los Alamos laboratory.

In 2017, biographies of two remarkable physicists were published. One, about Enrico Fermi, was titled, *The Last Man Who Knew Everything*. The other, *True Genius*, was about Richard Garwin. In 1949 at the age of 21, Garwin received a PhD from the University of Chicago. His supervisor was Enrico Fermi, who described him as "the only true genius he had ever met."[4] Fermi invited his young protégé to join him during summers in Los Alamos for continuing research there.

Fermi's Los Alamos office in the summer of 1951 was down the hall from Fred's. Fred recalled knocking timidly on Fermi's door. His self-described timidity might surprise friends who knew Fred to be bold and forthright about his ideas. But it also reflected Fred's sense of awe about Fermi and the other great physicists at the lab. Fermi invited him in. At the time Garwin, Fermi's young associate, was sharing his office. Garwin was present when Fred began his conversation with Fermi.

---

[4]William Broad, "Scientist at Work: Richard L. Garwin; Physicist and Rebel Is Bruised, Not Beaten," *The New York Times*, Nov. 16, 1999: F1.

"I'd like to talk to you a few minutes about the possibility of neutrino detection," Fred said.

"Well, tell me what's on your mind," Fermi replied warmly. Though bearing an Italian accent, his speech was famously clear and direct. He was a favored lecturer by students and colleagues.

"First off as to the source, I think that the bomb is best."

Understanding that Fred meant a fission bomb, Fermi agreed. "Yes, the bomb is the best source."

Pleased with the response, Fred then voiced his core concern. A presumptive detector would have to be enormous. "I don't know how to make such a detector," he said.

As Fermi considered the matter, Fred sat expectantly. Hoping for a technical suggestion or at least a word of encouragement, Fred received neither. According to Fred's recollection, Fermi simply said he couldn't think of how to make one either. Recounting the conversation, Fred said that "coming from the Master that was very crushing."[5]

Years later, in a 1995 congratulatory letter to Fred for winning the Nobel Prize, Garwin mentioned his own somewhat different memory of the conversation:

> I remember very clearly (which does not necessarily mean that my recollection is correct!) your coming to talk with Fermi about your proposal to do an experiment underground below a nuclear test exploded on the surface or on a tower, so that you could actually detect the neutrinos.... Probably it was Enrico who pointed out that it would probably be easier to get access to a nuclear reactor (and you could get much closer to it) and that a 300 MWt [Megawatt] reactor would produce a bomb's worth of neutrinos every few days.

Garwin ended his letter by again congratulating Fred for his ultimate success.

Whichever the correct version, after Fred's conversation with Fermi, he put the matter aside. Months later Fred mentioned his interest in the project to Clyde Cowan. Cowan joined the Los Alamos Laboratory in 1949 and, like Fred, was involved with weapons testing. When Fred suggested they

---

[5] Reines Nobel lecture.

work together on the neutrino problem Cowan readily agreed. Thus began their fruitful relationship. They began to design a large detector containing enough protons to presumably stop a few of the neutrinos emitted by a bomb. Fred described their rationale in an unpublished note:

> Our crude knowledge of the expected energy spectrum of neutrinos from a fission bomb suggested that the inverse beta decay reaction would occur several times in a several-ton detector located about 50 meters from the tower-based explosion of a 20-kiloton bomb.[6]

A 20-kiloton fission bomb! Shocking. The bomb would be comparable to the one that destroyed Hiroshima. In hindsight the proposal seems incomprehensible, yet Fred seemed dismissive of any concern.

> Anyone untutored in the effects of nuclear explosions would be deterred by the challenge of conducting an experiment so close to the bomb, but we knew otherwise from experience and pressed on.

By this time, bomb tests were being conducted with regularity. The safety and necessity of the testing program were still broadly accepted. Not until a decade later had much of the public come to oppose above-ground testing. Previously neither Enrico Fermi, Hans Bethe, nor other Los Alamos scientists voiced any ethical reservations about the Reines–Cowan plan to use a bomb.[7]

Norris Bradbury, Robert Oppenheimer's successor as Laboratory Director, granted approval. Without further elaboration about the size or nature of the explosion, Fred turned to the matter of detection. Recent findings had indicated that a weak flash of light would be emitted when gamma rays and charged particles passed through certain transparent liquids. Photomultiplier tubes could convert light into electrons and then intensify the electric signal. Placing a series of tubes at the edge of a detector could enable identification of various particles based on the emitted light signals. Fred referred to the overall mechanism — liquid medium, container that holds the liquid, and photomultipliers — as a scintillation device.

---

[6] Anderson: 11.

[7] *Ibid*: 10.

> The detector we dreamed up was a giant liquid scintillation device, which we dubbed "El Monstro." This was a daring extrapolation of experience with the newly born scintillation technique. The biggest detector until Cowan and I came along was only a liter or so in volume.

After designing the detection device, Fred and Clyde turned to the most formidable challenge: how to preserve the device and its effectiveness during the nuclear explosion. Clyde Cowan wrote about their plan years later:

> We would dig a shaft near 'ground zero' about 10 feet in diameter and about 150 feet deep. We would put in a tank, 10 feet in diameter and 75 feet long on end at the bottom of the shaft. We would then suspend our detector from the top of the tank, along with its recording apparatus, and back-fill [with dirt] the shaft above the tank.[8]

In advance of the explosion, vacuum pumps would evacuate the tank. When the countdown reached zero the suspension would be broken by a small explosive and the detector would drop through the tank. Next, according to Cowan's report:

> For about 2 seconds, the falling detector would be seeing the antineutrinos and recording the pulses from them while the earth shock [from the blast] passed harmlessly by, rattling the tank mightily but not disturbing our falling detector. When all was relatively quiet, the detector would reach the bottom of the tank, landing on a thick pile of foam rubber and feathers.

After a few days, when surface radioactivity had sufficiently dissipated, the Reines–Cowan team would return to the site, "dig down to the tank, recover the detector, and learn the truth about neutrinos." The figure below depicts the planned experiment. Cowan reiterated the contents of the report in a subsequent presentation at the U.S. Naval Academy. Using chalk and blackboard he deftly reproduced the diagram as he explained the plan.[9]

---

[8] Clyde L. Cowan, "Anatomy of an Experiment: An Account of the Discovery of the Neutrino," *Smithsonian Report for 1964*. Washington DC, Smithsonian Institution, 1965: 409–430.

[9] A video of the presentation is available on the Internet at https://www.youtube.com/watch?v=AYqEtm0X2Sc.

From an atomic explosion to a bed of feathers? The sequence of steps leading to the feathers resembled a Rube Goldberg contraption. The plan was approved by laboratory director Bradbury. With that came financial support. After the success of the Manhattan Project, scientists had gained a heroic aura, and scientific laboratories were often rewarded with generous post-war budgets.

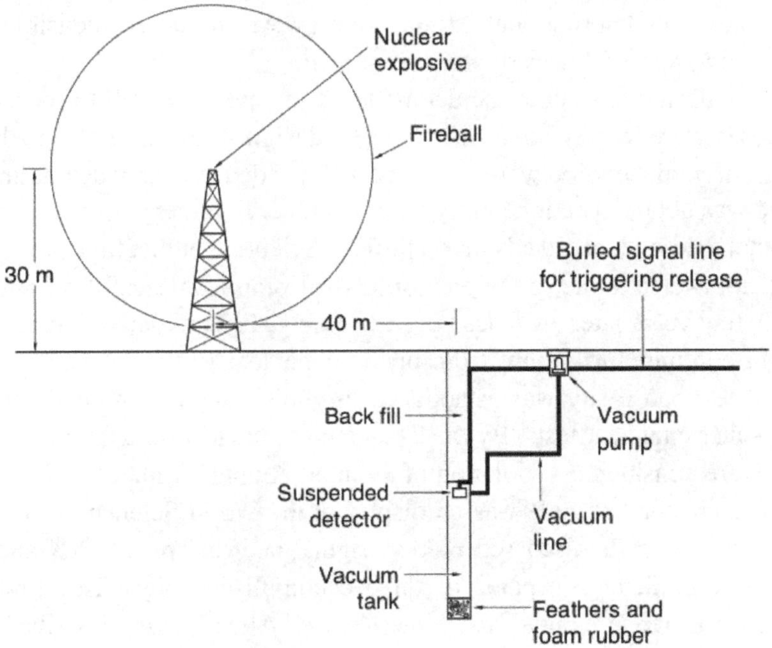

Detecting Neutrinos from a Nuclear Explosion.

Source: E.C. Anderson, *The Reines-Cowan Experiments, 1953–1956*: *Detecting the Poltergeist*, Los Alamos *Science*, No. 25, 1997.

The Reines–Cowan experiment quickly began to take shape. As Fred noted, Enrico Fermi and Hans Bethe eased the path by giving their "blessing [to] our idea."[10] The team built scaled-up models of scintillators filled with a mix of purified water and toluene. The detection of electrons and gamma rays proved to be unaffected by expanding a scintillator's size. Work also began on designing the large vacuum tank and the explosive mechanism that would release the detector. At the same time, the shaft that would hold

[10]Anderson: 13.

the tank beneath the atomic bomb was being drilled at the Nevada test site. Preparation for the explosion-based experiment proceeded into 1952.

Fred's idea to use an atomic bomb in his experiment was consistent with post-war attitudes. If a nuclear explosion were conducted for a beneficent purpose, it could be deemed virtuous. This notion was reinforced in 1953 by President Dwight Eisenhower at the United Nations General Assembly. In his Atoms for Peace speech he proposed the creation of an International Atomic Energy Agency — responsible for developing peaceful uses of atomic energy.[11]

Any hesitations about the detonation of an atomic bomb for peaceful purposes were largely about the project's design. Protection against high levels of radioactivity were assumed to be adequate and detonating a bomb was not deemed inherently objectionable. Judgments were based on whether the bomb could advance a project. Scientists at that time generally were not averse to the use of an atomic explosion for peaceful purposes.

A few years later attitudes began to change as cumulative radioactive fallout continued to mount from open-air nuclear testing. As a result, a partial test-ban treaty was established in 1963, which prohibited further atmospheric nuclear tests. By the 21st century, nuclear detonations of any kind were considered a violation of an international norm.

The changed attitude was on display at the World Science Festival in 2012 when experts discussed Fred's original neutrino plan. CNN anchor Bill Weir moderated a panel of three accomplished physicists: Francis Halzen, Lawrence Krauss, and Janet Conrad. After Halzen described the open-air explosion and the detector falling through a "hole in the ground," fellow panelists chuckled. Weir, feigning alarm, further injected humor by asking if a graduate student would be holding the descending detector. The panelists and audience roared with laughter. Halzen called the experiment a "crazy idea" and no one in the auditorium seemed to think otherwise.[12]

Ray Jayawardhana, the physicist author of *Neutrino Hunters*, wrote that the idea now sounds outlandish though he allowed that the plan was clever. When Fred shared his ideas with Enrico Fermi, Hans Bethe and

---

[11] Eisenhower, Atoms for Peace....

[12] The Elusive Neutrino and The Nature Of The Cosmos, World Science Festival, November 21, 2014, https://www.youtube.com/watch?v=CBfUHzkcaHQ.

others in 1951 they did not think his proposed use of the bomb was objectionable. Norms change over time and what was considered acceptable decades ago may seem inconceivable today.

Other "explosion-based" projects involving the bomb were far more provocative than Fred's. Project Orion proposed sending a spaceship to Mars and beyond. The program assessed differently sized vehicles for missions to various destinations. All would be powered by nuclear bombs. The project began in 1957 by a group of scientists led by physicists Ted Taylor and Freeman Dyson. They worked at the La Jolla, California, campus of General Atomic, a division of a company that sought to develop peaceful uses of atomic energy.

(Development of a small manned vehicle for deep-space exploration is currently underway. Ironically, the spacecraft is also named Orion. But it will be propelled by conventional chemical rockets, unrelated to nuclear explosives.)

One of the earlier Project Orion concepts included building a 4000-ton spaceship that would accommodate 20 or more individuals. The ship's size, 100 meters long and 50 meters wide, approximated the dimensions of a football field. Its interplanetary voyage was expected to run 3 to 5 years. No chemical fuel at that time could provide sufficient energy for the trip. With the advent of atomic and hydrogen bombs, such an effort became conceivable.

In one scenario, a launching would require 800 "low-yield" atomic bombs to overcome the Earth's gravitational pull and enable the ship to reach orbit. Located on the exterior of the ship, the bombs would detonate in quick succession, one every half-second. The energy released by each detonation would be less than half a kiloton. Other bombs with yields as high as 15 kilotons would remain attached to the craft for use during an extended flight.[13]

Even the smaller devices could cause grave harm if misdirected or if the vehicle and its passengers were inadequately shielded. Thus, every explosion had to be precisely calibrated as to strength and direction. Simulated tests were conducted with non-nuclear devices. The sound of rapid sequential firing during take-off prompted operators to refer to

[13] George Dyson, *Project Orion: The True Story of the Atomic Spaceship*. NY: Henry Holt, 2002: 2, 5, 182.

Project Orion as "Putt-Putt." Perhaps the most stunning determination was that a lengthy space expedition might require the ship to carry as many as 2600 atomic bombs.[14]

Fred Reines was not part of Project Orion though he had helped parent its main propulsion idea. In 1946 at the Los Alamos Laboratory, the mathematician Stanislaw Ulam posited the notion of powering a spacecraft with atomic bombs. Fred joined his effort to calculate how such a feat might work. In 1947 they co-authored an equation-filled memorandum on their formulations.[15]

A decade later, project development began. Fred, Ulam, and Conrad Longmire produced a then-classified report titled "Some Schemes for Nuclear Propulsion." It recapitulated the earlier calculations to produce a succession of "mild explosions" to propel a space vehicle.[16]

At the outset of Project Orion two dozen scientists, several of them alumni of the Manhattan Project, were consulted about its feasibility. George Dyson reviewed the leading role of his father, Freeman Dyson, and of other notables who had agreed to participate. The list included Hans Bethe, Stanislaw Ulam, Harris Mayer, Murray Gell-Mann, Robert Christy, Arthur Kantrowitz, and more. Only one of the scientists who visited the project's facility refused to engage further. As George Dyson wrote:

> Richard Feynman came down to General Atomic but declined the invitation to sign on. He called Orion "the pie in the sky." He did not believe the idea was technically impossible, he just did not wish to have anything more to do with secret projects or nuclear bombs. "Feynman was determined after Los Alamos never to do anything like that again," says Freeman [Dyson]. "He got too addicted when he was at Los Alamos and didn't want to be involved."[17]

---

[14] *Ibid*: 85.

[15] Duncan Geere, "Project Orion," *Medium*, March 3, 2014, https://medium.com/looking-up/project-orion-5c29e5477a99.

[16] Some Schemes for Nuclear Propulsion, http://publishing.cdlib.org/ucpressebooks/view?docId=ft9g50091s&chunk.id=d0e24994&toc.id=d0e24994&brand=ucpress.

[17] Dyson: 91–92.

Although Fred Reines was not among the participants, he wrote about it in an unpublished essay titled "The Limitations of Space Travel." He considered Orion an "ingenious approach," and that "it appears quite feasible to propel a space ship anywhere in our solar system by means of many small nuclear explosions."[18]

The initial willingness of so many scientists initially to participate in Project Orion reflects the marked difference in norms then from now. By today's standards such a project would be anathema. Yet in the post-war period, brilliant and morally sensitive individuals deemed the project worth pursuing.

During Project Orion's lifespan many technical challenges were identified and at least partially addressed. Among them were the shielding of passengers from radiation and the development of the pulsating system that would direct the ship. The shock-absorber of the Putt-Putt explosions was calibrated to cause no harm to the ship or its passengers. Still, the 1963 treaty that banned atmospheric testing of nuclear weapons effectively ended the project.

Many people now would agree with Francis Halzen that using an atomic bomb in a neutrino experiment is a crazy idea. Recent articles that inform the current generation about Project Orion are equally unsparing. One description likened it to "preposterous science fiction."[19]

Kelly Dickerson in *Business Week* wrote: "The simple absurdity of dropping a trail of nuclear bombs is a big reason Project Orion failed."[20] Few scientists in the 1950s considered Orion absurd. To be fair, one must be wary of applying current societal norms reflexively to judge past behaviors.

Science writer Ed Regis describes certain projects as "pathological technology." An example was Project Plowshare, which included the use of nuclear explosives for large nonmilitary projects. The explosions could

---

[18] Reines Collection.

[19] Armagh Observatory and Planetarium, October 1, 2010, https://armaghplanet.com/project-orion-the-incredible-nuclear-spacecraft.html.

[20] Kelly Dickerson, "Physicists Once Tried to Build a Spaceship Propelled by Nuclear Bombs, *Business Insider*, June 11, 2015, http://www.businessinsider.com/project-orion-nuclear-bomb-propelled-spaceships-2015-6.

create new sea-level canals and flatten large rock formations that obstructed highway construction. Regis wrote:

> It is easy for us today to recognize Project Plowshare for the lunatic folly that it was. But it didn't seem so at the time — at least not to some of the nation's best and brightest scientists.[21]

Describing the proposals to use nuclear explosives as crazy, absurd, or lunatic, may appear warranted when current standards are applied. But scientists who championed the notion were not the only people who gave support to Plowshare. It was part of the Atoms for Peace initiative introduced by President Eisenhower. The initiative was included in the 1954 Atomic Energy Act, whose passage was supported by members of both the Democratic and Republican parties. The new law drew endorsements throughout American society — from industry, labor unions, scientists, engineers, and public interest groups. The idea of using nuclear explosions for peaceful purposes was undisputedly popular.[22]

Examples abound of cultural norms during one historical period being later viewed as reprehensible. This is true of slavery, denial of women's suffrage, research on unwitting human subjects. Unforgiving criticism today of earlier nuclear-based proposals reflects a change in societal attitudes. Contemporary norms are subject to future changes as well. In 2017, 122 nations signed an international agreement to ban all "nuclear weapons or other nuclear explosive devices."[23] Aimed at eliminating the possibility of nuclear war, the treaty would also seemingly end options to use an atomic explosion for peaceful purposes. Skeptics consider the agreement naïve and oppose the apparent trend toward global acceptance. Neither the United States nor other nuclear powers were signatories.

Of course, current standards do not prohibit all things nuclear. Society still largely accepts the use of nuclear reactors to produce energy.

---

[21] Ed Regis, *Monsters: The Hindenburg Disaster and the Birth of Pathological Technology*. NY: Basic Books, 2015.

[22] Richard G. Hewlett and Jack M. Holl, *Atoms for Peace and War, 1953–1961: Eisenhower and the Atomic Energy Commission*. Berkeley, CA, University of California Press, 1989: 136.

[23] Treaty on the Prohibition of Nuclear Weapons, http://undocs.org/A/CONF.229/2017/8.

Nuclear-powered submarines remain a mainstay of the U.S. Navy. Still, a norm conceivably could emerge that excludes nuclear-based power of any sort. Will future generations condemn today's nuclear plants and nuclear-powered submarines as crazy and ludicrous? To raise the question is itself a plea for humility in a search for answers.

Toward the end of 1952, Reines and Cowan reported progress on their bomb-based experiment, at a Los Alamos seminar for the laboratory's Physics Division. Their presentation prompted an unexpected response from Jerome Kellogg, the division leader. He suggested that they consider using a fission reactor instead of a bomb as a source of neutrinos. His proposal was not couched in moral terms but rather as a matter of practicality. Comments from the audience of physicists largely ignored Kellogg's idea. They supported continuing with a bomb-based approach.

Later that evening Reines phoned Cowan to discuss Kellogg's suggestion. True, a reactor, in a few seconds or even a few hours, yields thousands of times fewer neutrinos than a bomb disgorges in a fraction of a second. This would make it harder to identify its neutrino signal amid the numerous signals from cosmic and other sources. But unlike an explosive device, a reactor would permit reuse of the detection apparatus.[24]

During their conversation Fred and Clyde decided to reconsider their approach to detection. Despite the smaller flux of neutrinos, the use of a reactor suddenly seemed feasible. "We immediately altered our plans and the next morning met with Kellogg to cancel our bomb preparations and arrange to develop and build a detector suitable for the Hanford reactor [in Washington state]."[25]

Fred and Clyde soon came up with a distinctive way to identify a neutrino amid background events. They figured that a unique two-step process could be the giveaway to identification. Each step yields a light signal that could be picked up by the detector.

The first signal is a pulse of light caused by gamma rays resulting from the annihilation of positrons (antielectrons) during inverse beta

---

[24] Christine Sutton, "Ghosts in the Machine," *CERN Courier*, July 8, 2016, http://cerncourier.com/cws/article/cern/65501.

[25] Frederick Reines, "The Early Days of Experimental Neutrino Physics," *Science*, January 3, 1979.

decay. The decay also produces a neutron. The slower moving neutron is then captured by the nucleus of a cadmium atom, which emits gamma rays that prompt a second light signal. The time between the two pulses is only a few microseconds. The "delayed coincidence" provides the distinctive signature of inverse beta decay. Fermi's formulation of the decay process would require that the elusive particle must have been present. Therein could be the key to a neutrino's detection.[26]

By the end of 1952, the effort to use a reactor rather than a bomb was underway.

[26] Anderson: 3–4.

Chapter 7

Hanford and Savannah River

When Reines and Cowan began to consider a nuclear reactor for the neutrino experiment the obvious choice was at Hanford in south-central Washington. They had previously spent time there. Hanford was established during World War II as the main site for producing plutonium to fuel atomic bombs. The 586-square-mile area was flat, dry, and dotted with shrub vegetation. After the war, Hanford's reactors continued to churn out plutonium for the growing U.S. nuclear arsenal.

Detection of neutrinos from a reactor would differ markedly from recording those from a bomb. Fred was especially pleased by Enrico Fermi's response to the change of plans.

> October 8, 1952
> Dear Fred:
>
> Thank you for your letter of October 4th by Clyde Cowan and yourself. I was very interested in your new plan for the detection of the neutrino. Certainly your new method should be much simpler to carry out and have the great advantage that the measurement can be repeated any number of times. I shall be very interested in seeing how your 10 cubic feet scintillation counter is going to work, but I do not know of any reason why it should not.
>
> Good luck.
> Sincerely yours,
> Enrico[1]

---

[1] Reines Collection.

Fermi's theory of inverse beta decay was central to Reines' thinking. Such decay involved an occasional interaction between a proton and a presumed neutrino.

Reines and Cowan's previous work in anticipation of using a bomb explosion had itself been instructive. Prior to discarding the plan to seek bomb-generated neutrinos, Reines and Cowan had built their liquid scintillation detector. "El Monstro" was a tank about three cubic feet in volume, which contained 260 gallons of liquid. Ultimately it was never used in a neutrino experiment, but it nevertheless offered valuable lessons. Tests showed that electrons and gamma rays would generate scintillation patterns, whatever the size of the detector.

With attention now on a test involving a nuclear reactor, Reines and Cowan brought on other colleagues to help with their "Project Poltergeist." The eight-member team spent the winter in Los Alamos preparing for the test. They assembled equipment and built a detector, which they named "Herr Auge" (German for "Mr. Eye"). The detector would hold 80 gallons of scintillator. One of the team members, E.C. Anderson, later described their venture to Hanford early in 1953, when "the Project Poltergeist team packed up Herr Auge [and] numerous electronics and barrels of liquid scintillator, and set out for the new plutonium-producing reactor."[2]

The equipment filled two trucks for the 1250 mile trip from Los Alamos to Hanford. Upon arrival, the detector was brought into the building that housed a large fission reactor. The team placed the Auge detector close to the reactor. After Herr Auge was filled with scintillator liquid it was surrounded with a shield comprising layers of paraffin, boron, and lead. The electronic gear was set up to detect "the telltale delayed-coincidence signal from inverse beta decay," Anderson wrote. The shielding was intended to block background neutrons and gamma rays from reaching the detector and interfering with the desired signal.

The aim was to detect the two predicted sequential light signals. The first was caused by gamma rays resulting from the rare interaction of a neutrino with a proton. A positron from the neutrino capture collides with

---

[2] Anderson: 8.

an electron in the scintillator liquid. The resulting annihilation of the oppositely charged particles produced gamma rays and a visible light pulse. A second pulse would occur when a slower moving neutron was caught by an atom of cadmium. (Cadmium had been added to the scintillator to induce the reaction.) As Fred's friend Harris Mayer further observed:

> The two pulses recorded by photomultipliers were fed into delayed coincidence discriminating electronics. An acceptable signal had amplitude and time delay easily distinguishable from the very many randomly occurring pulses due to background radiation.[3]

If the second pulse appeared within 9 microseconds of the first pulse they were recorded as a delayed coincidence — a pattern signifying the presence of a neutrino. Trillions of neutrinos would be passing through the detector every second. A delayed coincidence signal could be expected about once or twice every 10 minutes.

On the eve of the experiment, Fred Reines wrote a letter to his mother, my aunt Gussie. (His father, Israel Reines, had died 3 years earlier.) The letter, dated March 8, 1953, revealed Fred's excitement as he sought fulfillment of his long-held ambition, to be a physicist "extraordinary." An excerpt:

> Dear Mom:
>
> As you can tell by the letterhead I am in the State of Washington. For the past year or so, a friend of mine, Clyde Cowan and I have been working on a wild idea. We wanted to detect a particle called a neutrino which, though it plays an important part in the nature of things, has such characteristics given it by theoretical considerations that direct detection has been pronounced "absolutely impossible" by many leading scientists. If it could be detected, that would be a great discovery — if it were found not to be there, that would be an even greater one. In either case it's the sort of work they give Nobel Prizes for.
>
> Working together, "practically inside each other's heads," Clyde and I developed a new technique for the job, assembled a crackerjack

---

[3] Harris L. Mayer, "Catching the Free Neutrino," unpublished manuscript, July 14, 2002.

group, [and] got the authorization from the Lab to try it. Now we are here at the AEC's [Atomic Energy Commission] Hanford Engineering Works trying our luck. In the course of developing the technique we have built a detector which is many, many times more sensitive than any ever built before and have opened new fields of physics and biophysics with it....

Of course we are excited by these developments, especially as we reach now for the great prize, the neutrino. We have submitted a few letters and a paper to scientific journals describing these developments and we expect quite a splash when they hit print.

If our neutrino experiment works as planned, the Reines name will be long remembered for it. Our group here has been working 7 days a week, 15 hours a day, so keen is their interest, so great their excitement. Little did you and Pop know when you worked so hard at the Washington Dairy that one day you would get the reward which is so close at hand. Must not get too optimistic, but gosh things look good.

Sylvia and the children are anxiously waiting at home for me to finish here and return to them. I'm looking forward to the same thing.

I've never worked so hard in my life but Oh something tells me that this is worth the utmost I can give and then some.

I hope you are well, Mom, and can forgive me for not writing for such a long time. Regards to all.

Love, your son,
Fred

Gussie Reines later had copies of the letter typed, which she shared with several relatives.

Soon after the experiment was underway, Fred's excitement and optimism began to fade. The initial findings were troubling. Whenever measured, the delayed coincidence counts were far greater than anticipated.

"The delayed coincidence background, present whether or not the reactor was on, was about 5 counts per minute, many times higher than the expected signal rate," Anderson wrote.

For several months, the team stacked and restacked the shielding and used various recipes for the liquid scintillator.... Then they would set the electronics and listen for the characteristic double clicks that would

The Hanford Team, 1953. Reines, 3rd from right; Cowan far right, LANL.

Source: E.C. Anderson, "The Reines-Cowan Experiments, 1953–1956: *Detecting the Poltergeist*, Los Alamos *Science*, No. 25, 1997.

"Herr Auge," the Hanford Neutrino Detector. It contained 80 gallons of scintillator liquid surrounded by 90 photomultiplier tubes.

Source: E.C. Anderson, "The Reines-Cowan Experiments, 1953–1956: *Detecting the Poltergeist*, Los Alamos *Science*, No. 25, 1997."

Reines in detector area at Hanford.

Source: E.C. Anderson, "The Reines-Cowan Experiments, 1953–1956: *Detecting the Poltergeist*, Los Alamos *Science*, No. 25, 1997.

> accompany detection of the inverse beta decay. Despite the exhausting work, the results were not definitive.[4]

The group began to recognize that their shielding was inadequate. Cosmic rays apparently were reaching the detector and mimicking signals that may also have been arriving from neutrino interactions, and the entire shielding and detection arrangement needed to be reassessed. Their earlier hopes remained unfulfilled and in the summer the group returned to Los Alamos. As Reines continued to pore over the Hanford data he realized that for a six-hour period, when the reactor was on, the data showed a slight rise in what looked like inverse beta decay. The cause presumably

---

[4] Anderson: 8.

was from neutrino interactions. Cowan subsequently recounted their pleasure with the belated indication, while recognizing that the observation was only suggestive at best.

> We did record neutrino-like signals but the cosmic rays with their neutron secondaries [i.e. associated neutrons] generated in our shields were 10 times more abundant than were the neutrino signals. We felt we had the neutrinos by the coattails, but our evidence would not stand up in court.[5]

When Reines and Cowan reported on their experiment to the American Physical Society, their optimism overshadowed Cowan's not-stand-up-in-court assessment. In September, a *New York Times* headline about their presentation conveyed little doubt: "Atom Team Sees 'Ghost' Neutrino: Scientists Believe They Have Observed Something Theory Says Has to Exist."[6]

Other headlines around the country were even more definitive:

> "AEC Pair Prove Existence of Elusive Neutrino," *Long Beach Independent* (California).
>
> "AEC Scientists Say Neutrino's Evidence Confirmed by Experiments," *Aiken Standard and Review* (South Carolina).
>
> "A-Experts See New Particle of Nucleus," *Corpus Christi Times* (Texas).
>
> "Mass-Less Weight-Less Neutrino 'Spotted'," *Bridgeport Post* (Connecticut).[7]

A press release in *Science Service* was no less declarative. Headlined, "Neutrino Now Detected," it stated that "the textbooks that now say that the neutrino has never been detected will have to be revised. [The] poltergeist of modern physics … has been caught."[8]

In November, Reines and Cowan described their experiment in a leading scientific journal, *Physical Review*. The unambiguous title was,

---

[5] *Ibid*.

[6] *The New York Times*, September 4, 1953.

[7] Images of the newspaper pages are at Newspaperarchive, https://newspaperarchive.com/tags/1953-neutrino/.

[8] *The Science News-Letter* (1954), 65, 7: 99.

"Detection of the Free Neutrino." Their introductory passage indicated that they would seek additional confirmation but that they had probably achieved their goal:

> An experiment has been performed to detect the free neutrino. It appears probable that this aim has been accomplished although further confirmatory work is in progress.[9]

Decades later, physicist Robert Arns noted the large gap between the numerous press reports and the actual findings. Even Fred and Clyde's *Physical Review* article failed to reflect the skepticism they expressed in private. In a laboratory notebook Cowan had written that slight variations in electron counts seemed attributable to neutrinos, but "we realize that we have not proved anything." Reines later revealed that in the immediate aftermath of the Hanford experiment, he and Cowan felt they had not identified any neutrinos. Only after later reviewing the data did they think they had found "a hint of a signal."[10]

After Reines and Cowan returned to Los Alamos they began to consider next steps. Encouraged by Laboratory Director Norris Bradbury, they established a new team exclusively to hunt down neutrinos. The group set about designing a markedly different detection system. The fundamental challenge was to differentiate between neutrino telltale signals coming from the reactor and confounding background signals. For this they would need a larger and more sensitive detector.

Detection at Hanford had depended on the signals from the back-to-back pulses a few microseconds apart, a delayed coincidence. The new group planned to add a spatial component to the identification efforts. The presumption was that the location of the signals could reinforce evidence based on the timing of the pulses. They designed a broad 10-foot-tall structure divided into five levels. The configuration resembled a club sandwich. The first, third, and fifth levels (the bread) were large plastic tanks of liquid scintillator. The second and fourth levels (the meat) were

---

[9] Frederick Reines and Clyde L. Cowan, "Detection of the Free Neutrino," *Physical Review*, November 1, 1953: 830.

[10] Robert G. Arns, "Detecting the Neutrino," *Physics in Perspective*, Birkhauser Verlag, Basel, 2001: 314–334; Reines, "The Early Days of Experimental Neutrino Physics": 203.

flatter tanks containing water. The scintillator in which a light signal initially appears could help identify its source — whether from the reactor or elsewhere.

As Harris Mayer wrote, the idea behind the new detector was simple: If separate scintillation tanks were placed one on top of another, a cosmic ray would pass through all of them, giving a counterfeit signal in each. An interacting neutrino could not do this. Its interaction in one of the scintillators would be completed and there would be no further signals in any other scintillator.[11]

The small cross-section (low probability) of a neutrino interaction prompted the building of a much larger detector than the one at Hanford. The Hanford detector had held 80 gallons of scintillator liquid. The detector tanks in the "club sandwich" held a total of more than 1100 gallons. The greater volume could increase the chances of recording a neutrino interaction. In addition, some 100 photomultiplier tubes encircled each scintillator tank. The tubes could intensify electric signals in the liquid to further enhance particle detectability.

Year-long construction and testing of the new detector and the associated electronics continued into the fall of 1955. During that time, John Wheeler, a distinguished physicist and supporter of the project, helped arrange for the location of the forthcoming experiment. It would be conducted outside the new reactor at the Savannah River site in South Carolina.

The 310 square-mile Savannah River site was established in 1950 as part of the nation's nuclear weapons program. As at Hanford, it produced plutonium. It also produced tritium and other nuclear-related materials. The newest Savannah River reactor could yield a larger flux of neutrinos than any other existing reactor.

In November, Reines and Cowan oversaw the packing of the equipment they had developed during the past year at Los Alamos. The combination of tanks, scintillator fluid, cables, time delay analyzers, amplifiers, and other electronic equipment filled five large trucks. "Translating our ideas into hardware," Fred had said, "was a formidable task." Transporting the weighty and delicate hardware over 1600 miles of

[11] Harris L. Mayer, "Catching the Free Neutrino," unpublished manuscript, July 14, 2002.

roadway was also daunting. One of the trucks was a flatbed trailer that carried the three large steel cylindrical containers filled with liquid for delivery to the detector tanks. Described as the "tank farm," their huge bulk would require them to remain outside the reactor building, while their assorted tubes and cables supplied the detector tanks inside.

Betty Cowan, Clyde's wife, in a later interview with a South Carolina newspaper described the journey as rugged. The Reines family including their two children and the Cowans with their three children, traveled in separate cars. They were followed by a car carrying other members of the research team. Behind them were the five truckloads of testing equipment. To minimize jolts to the equipment, the caravan crept along slowly. The trip took ten days. Betty elaborated:

> Each family took turns driving well ahead of the convoy, so that we could find a good motel and settle down early in the evening. But it never worked out that way. Somewhere along the route we would come to a viaduct with a clearance of only 11 feet 8 inches, when our tallest truck measured 13 feet 11 inches. Then we had to turn around in a hurry and find a detour. We posted signs on telephone poles to warn the trucks — and more than once we were stopped by a highway patrolman who wanted to know what we were up to.
>
> Late at night we would arrive at a motel and ask for accommodations for two families, several single men, and parking space for five trucks. They thought we were crazy.[12]

Besides apprehension about jostling the apparatus, the weather also was worrisome. As Reines recalled, "We were so concerned that the cold weather would cause precipitation of the scintillator solute that we wrapped electrical heaters around our specially constructed storage tanks."[13]

When the team arrived at the Savannah River site they parked the trucks near the reactor building. Then they drove to nearby Aiken, South Carolina, to settle into their temporary homes. In the days ahead, the

---

[12] Kathryn Lawrence, "The Life of a Scientist's Wife," *The State* (Aiken), April 22, 1956. Forty years later a videoed recollection by Betty differed somewhat, though not materially, from her account to the reporter after their arrival at Savannah River, https://www.youtube.com/watch?v=au3SCXknthM.

[13] Reines, "The Early Days of Experimental Neutrino Physics": 14.

trucks were unpacked and the cargo moved into and around the reactor building. The reactor was located underground, and the detector was placed adjacent to it. Nearly 40 feet of earth and concrete above them would reduce confounding effects of cosmic ray signals.[14]

The large three-tank farm remained outside the building where it would be connected by piping to the detector inside and four-stories down. The remaining underground space was minimal and much of the electronics therefore had to be placed in a tractor-trailer outside the building. The installation took 3 months to complete. Connections were checked and rechecked until the system was ready to be turned on.

The Savannah River Team, 1955. Cowan standing far left; Reines 4th from left, LANL. *Source*: E.C. Anderson, "The Reines-Cowan Experiments, 1953–1956: *Detecting the Poltergeist*, Los Alamos *Science*, No. 25, 1997.

---

[14] Arns: 323.

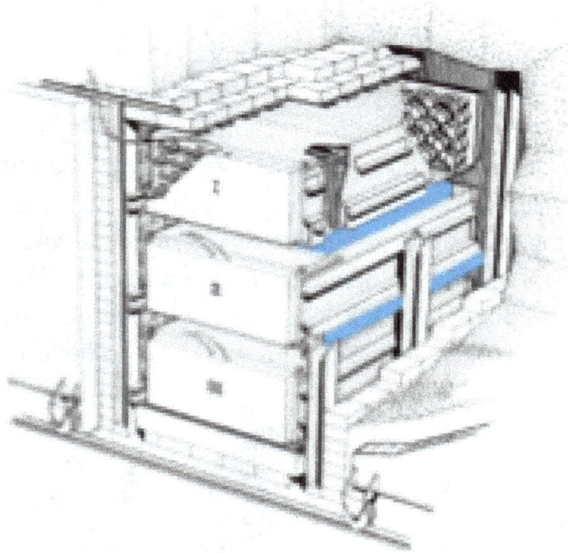

The Savannah River Neutrino Detector. Two flat tanks (blue) each contained 50 gallons of water. They lay between three tanks that each contained about 370 gallons of scintillation liquid.

Source: E.C. Anderson, "The Reines-Cowan Experiments, 1953–1956: *Detecting the Poltergeist*, Los Alamos *Science*, No. 25, 1997.

As Clyde Cowan later recalled:

> The new year [1956] found our detector installed near the great reactor, with pipes and bundles of wires and coaxial cables running to the laboratory and tank farm trailers parked outside. Calibrations were undertaken ... and the cosmic rays and backgrounds were measured in the myriad different forms they assume in such equipment. By early spring we felt that we were ready for our quarry.[15]

The team was excited, but also apprehensive. The Hanford experience was a reminder that results could fall short of expectations. What if the new experiment failed to dispel uncertainties associated with the Hanford results? Beyond their own disappointment, Reines and Cowan might be seen as having wasted time and resources. Would they be accused of leading a costly misadventure?

[15] Cowan, "Anatomy of an Experiment": 424.

At the end of 1953, amid the news about the earlier Reines–Cowan experiment, a seemingly unrelated event had taken place near Schenectady, New York. Irving Langmuir, a 1932 Nobel laureate in chemistry, made a presentation at a laboratory affiliated with the General Electric company. He titled his subject, "Pathological Science."

He recounted instances when findings by reputable scientists were greeted with excitement but later proved to be invalid. He did not include hoaxes or other intentional efforts to deceive. Rather, as he put it: "To me, the thing is extremely interesting, that men, perfectly honest, enthusiastic over their work, can so completely fool themselves."[16]

A transcript of Langmuir's presentation was widely circulated. Among his examples was the reported discovery of N-rays in 1903 by French physicist Prosper-René Blondlot. Unlike X-rays, which had been discovered 8 years earlier, N-rays allegedly could not penetrate water, though they could pass through aluminum and other metals. Blondlot was a respected member of the French Academy of Sciences. Langmuir noted that other scientists published "lots of papers" on N-rays and initially about half of them confirmed their existence. As time passed, the notion of N-rays proved untenable and Blondlot's reputation was tarnished.

A similar example involved so-called "mitogenetic" radiation. Beginning in 1923, Russian biologist Alexander Gurwitsch and others published papers on mitogenetic rays. Emanating from the cells of plants and animals, they were said to stimulate the growth of nearby cells. According to Langmuir, numerous experimental efforts failed to confirm the existence of mitogenetic rays. Suppositions about their existence were largely discredited even though a few scientists persisted in holding that they were real.

A third example of pathological science, the Davis–Barnes Effect, involved Langmuir himself. In 1929, Bergen Davis, a physicist at Columbia University, reported on the effects on a helium atom when targeted by a beam of electrons. The results were based on counting electrons that elicited scintillations visible through a microscope; Davis invited a skeptical Langmuir to his laboratory to examine the experiment's set-up.

---

[16] I. Langmuir, "Pathological Science," December 18, 1953: 9, http://galileo.phys.virginia.edu/~rjh2j/misc/Langmuir.pdf.

After hours of review Langmuir still believed that the experiment lacked merit. At his 1953 presentation, Langmuir recounted his response to Davis:

> I wrote him a 22-page letter giving all our data and showing really that the whole approach to the thing was wrong; that he was counting hallucinations, which I find is common among people who work with scintillations if they count for too long.

Davis issued no rebuttal and the issue just faded away. Two decades later, as a result of Langmuir's 1953 presentation, the erroneous Davis–Barnes Effect became better known. It was one of Langmuir's examples of behavior that was more common than generally realized. His presentation had been a cautionary message about unwarranted exuberance. From the transcript of his presentation:

> The characteristics of this Davis–Barnes experiment and the N-rays and the mitogenetic rays, they have things in common. These are cases where there is no dishonesty involved but where people are tricked into false results by a lack of understanding about what human beings can do to themselves in the way of being led astray by subjective effects, wishful thinking or threshold interactions. These are examples of pathological science. These are things that attracted a great deal of attention. Usually hundreds of papers have been published upon them. Sometimes they have lasted for fifteen or twenty years and then they gradually die away.[17]

Although Reines and Cowan may have been unaware of Langmuir's talk on pathological science, they presumably were aware of the risks of exaggerated claims of success. Yet media affirmations of success at Hanford had been based in part on the positive tenor of their public presentations. In some conversations with colleagues, however, Fred acknowledged that the Hanford findings had been inconclusive. Their initial public reports implied greater certainty than he and Cowan had expressed in private. Now, 3 years later, at Savannah River, the anticipated grand moment had arrived.

---

[17] I. Langmuir, "Characteristic Symptoms of Pathological Science," https://www.cs.princeton.edu/~ken/Langmuir/langB.htm#Characteristic%20Symptoms.

The starter switch was turned on. And then — nothing. No signals. Nothing. Fred was seized with a nightmare possibility. "Maybe there were no neutrinos," he thought, "or maybe they existed but were unstable and did not reach our detector." Had he and Cowan spent the years of preparation in vain? He wondered at that moment if the suggestive results from the Hanford experiment had been just a "flight of fancy."[18] Was he guilty of what Langmuir called pathological science?

Detector operators reached for the dials. They anxiously twisted and spun them until, to their immense relief, the detector responded. After the retuning, signs of cosmic activity began to appear. During the ensuing weeks they continued to calibrate and test and by mid-April they believed they were receiving confirmatory signals of neutrinos.

Reines and Cowan, Savannah River Laboratory, 1956, LANL, Reines Collection.

Further testing included surrounding the detector with extra shielding, which reduced the number of reactor-generated coincidences. The frequency of true delayed coincidences remained the same, confirming the

[18] Reines, "The Early Days of Experimental Neutrino Physics": 14.

presence of rare but decipherable neutrino interactions. As they monitored the detector for another 2 months, the observed events approximated the number expected. On June 14, 1956, they sent a telegram to Wolfgang Pauli in Zurich. He was the man who in 1930 had theorized the existence of neutrinos, but believed they could never be detected: "We are happy to inform you that we have definitely detected neutrinos from fission fragments by observing inverse beta decay of protons."[19]

The telegram reached Pauli at CERN, the European accelerator research site in Geneva that had been established 2 years earlier. He interrupted the proceedings of the conference to read the message to the audience, and Fred later learned that Pauli and some friends had consumed a case of champagne in celebration. Pauli also telegrammed a response to Reines and Cowan, though they never received it. A copy was found in the Pauli archives. Avoiding reference to his supposition that neutrinos would be undetectable Pauli simply quipped: "Everything comes to him who knows how to wait."[20]

Fred's 1953 letter to his mother anticipated the excitement that detection of the neutrino would bring. "It's the sort of work they give the Nobel Prize for," he had written. For many years after his success in 1956 at Savannah River that thought remained with him.

[19] Reines Collection.

[20] Jayawardhana: 72.

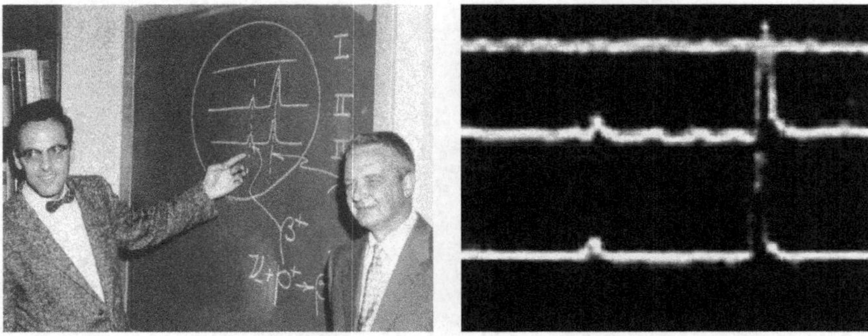

Reines and Cowan explain the neutrino signature: "delayed coincidence." A positron blip signal is followed by a taller signal of neutron capture gamma rays, Reines Collection.

Chapter 8

Clyde Cowan

The close association between Clyde Cowan and Fred Reines ended after their 5-year collaboration on the neutrino experiment. Cowan left Los Alamos in 1957 to join the faculty at George Washington University in Washington, D.C. The following year he transferred to nearby Catholic University of America where he remained a physics professor until his death in 1974. Unlike Fred, whose research on neutrinos continued throughout his career, Clyde's research interests at Catholic University were focused elsewhere. His path to Los Alamos and their joint discovery also had differed from Fred's.

Clyde was born in Detroit on December 6, 1919. His family moved to St. Louis where he attended public schools. In 1940 he received a bachelor's degree in chemical engineering from the Missouri School of Mines. There he was in the Reserve Officers Training Corps (ROTC) and after America entered the war in 1941, he joined the Army Air Force as a second lieutenant. He was stationed at the newly established branch of the Massachusetts Institute of Technology Radiation Laboratory in Great Malvern, England, where he helped advance radar technology. Applying his skills in chemical engineering, he also built an experimental cleaning unit in case of a gas attack. He met Betty Dunham in October 1942, and 4 months later they were married. Betty believed that she was the first British bride of an American soldier.

After the war ended in 1945, Clyde received permission from the army to return to the United States via civilian transportation. Betty's father, Henry Dunham, was in the paper business and he secured a spot for the couple on a small freighter carrying paper goods. Clyde, Betty, and their 18-month-old daughter, Elizabeth, left England for a 3-week trans-Atlantic voyage. As she recalled in a video-taped interview 50 years later:

> We landed in Montreal. We came down through Detroit, though I had a hard time getting across the border. The Americans had all kinds of restrictions because of the war and apparently, I came in the "wrong" way. Eventually they relented.[1]

Clyde was assigned to Wright Patterson Air Base in Ohio and they lived in Dayton for the next year. He left active duty in 1946. With funds for education available through the GI Bill, he began graduate studies at Washington University in St. Louis. After receiving his PhD in physics in 1949, he joined the staff at the Los Alamos National Laboratory. He met Fred soon afterwards and initially they worked together on some nuclear military projects. Then in 1951 they began as a team to pursue Fred's earlier interest in neutrino detection. The highlight of the Reines–Cowan experience occurred in South Carolina. As described in Chapter 7, their two families had moved to Aiken in late 1955 and the scientists would conduct their experiment at the nearby Savannah River research site. The "wow" moment came on June 3, 1956 when Reines and Cowan, abetted by other members of the Los Alamos team, detected the neutrino. The confirmative findings were exciting. Still, in 1957 Clyde decided to move on.

After arriving at Catholic University in 1958, Cowan began a project that remained his primary research effort for the next 15 years. While engaged in some neutrino-related investigations, he mainly focused on a search for the source of muons. His experiments were largely based on muon interactions observed in a spark chamber detector located in the sub-basement of a university building. A 1969 paper included a description of the latest findings in his effort. An excerpt:

> We have engaged for some time in the investigation of reactions at sea level produced by neutral components of the cosmic radiation which

---

[1] Clyde Cowan, Catholic Scientist, https://www.youtube.com/watch?v=au3SCXknthM; Elizabeth Cowan Riordon interview, 2019.

result in at least one muon of either sign and with energy lying between a few tens and a few hundreds of MeV [million electron volts]…. The charge ratio of [negative to positive muons is] obtained from 76 events collected in 2,475 hours of run time.[2]

Hall Crannell was familiar with Cowan's project. He had joined the Catholic University physics faculty in 1967 after completing his PhD and post-doctoral studies at Stanford University. He considered Cowan a brilliant experimentalist. In 2019, at age 82 and a professor emeritus, he described Cowan's muon work to me.

Actually, he never found the answers he was looking for. He saw muons. He could look for when a muon went through the spark chamber. His group could take pictures of where the sparks were, and they could use that to project where the muon came from back in the sky.

So they would take these pictures and then students would scan the pictures with a device that gave the coordinates of what they saw. After looking at a lot of muons Clyde's group would find more of them than would be expected statistically to come from some point in the sky. They would then write up their findings.

Being suspicious about their findings they would build a bigger experiment to get better data. When they would get better data they would find that the area was not a source. Then they would see another area that looked like it might have a surplus. So this went through [multiple] iterations.

Being pretty much involved in statistics myself, I was always doubtful that they were seeing anything substantive. You know, when you look at a thousand different points some of them occasionally look bigger than you might have expected. So really, he didn't determine anything from that experimental effort.[3]

Among the funders of Cowan's project were, variously, the Atomic Energy Commission, the National Science Foundation, and the Navy and

[2] Clyde Cowan, *et al.* The Production of Low Energy Muons by Neutral Components of the Cosmic Radiation. Presented at the Eleventh International Conference on Cosmic Rays, Budapest, 1969.

[3] Hall Crannell interview, 2019.

Air Force. I wondered how he managed to continue to get funding. Crannell responded:

> Oh, it's very exciting. The funders would say, Look, he's done this big experiment. He's a good experimentalist. He's got lots of data that he shares with everybody. He points out that he's seen this surplus of events from some particular direction. This could be a remarkable event if it's for real, so now he wants to investigate it by doing a bigger and better experiment. Science is full of things like that.[4]

During Cowan's years at Catholic University, Fred Reines periodically visited government officials in Washington regarding his ongoing neutrino experiments. The Atomic Energy Commission continued to be a principal funding agency for his work. While in Washington Fred would occasionally meet with Clyde. Hall Crannell also got to know Fred and he recognized their contrasting personalities. Reines sparkled with enthusiasm. He was a charismatic extrovert and Cowan was more sedate and restrained, less outwardly expressive.

While speaking with Crannell I referred to a lecture that Cowan had given at the Naval Academy on the discovery of the neutrino. Available on YouTube, Cowan seemed laid back. He spoke without notes while unfolding a compelling story clearly and calmly.[5] Fred's manner of delivery, as evidenced by taped lectures elsewhere, was more emotive. He displayed a greater sense of excitement.

"Yes indeed," Crannell agreed, and then offered an additional perspective after calling both men fine experimentalists:

> I always thought they must have made a great team. Clyde was a genius in the laboratory although not a particularly good manager. Whereas Fred was really a clever manager-type. So they made a wonderful team.[6]

Another of their differences lay with their attitudes about religion. Reines was Jewish, but not religious. Cowan was a devout Catholic who

---

[4] *Ibid.*

[5] The Neutrino, with Clyde L. Cowan https://www.youtube.com/watch?v=AYqEtm0X2Sc.

[6] Crannell interview, 2019.

fully embraced the teachings of the Church. Long after Clyde's passing, in 1996 his widow Betty and their son George were interviewed on a program sponsored by the Knights of Columbus, a Catholic fraternal organization. The program was titled, "Clyde Cowan, a Catholic Scientist." The interviewer, Francis Kelly, a physicist, had been a student of Cowan's at Catholic University. He remembered Clyde as a caring mentor.

Betty mentioned that Clyde had been a member of the Knights of Columbus and lectured about the creation of the world from a Catholic's perspective. He was, she said, "very Catholic." George also spoke of his father's deep faith:

> I remember him saying the more he learned and the more mysteries he unraveled, the deeper his faith became. Many people have asked, How can a scientist also have faith? Yet my father found that the science enhanced his faith.[7]

In 2019 George reaffirmed to me his earlier characterization:

> The more Clyde learned the more beauty he said he could see in what he was doing, and it just reinforced his faith. I think it was a rather nifty aspect of my father. He could have just reduced his findings to a series of equations to understanding relationships and what-not. But he didn't succumb to that.[8]

In the Knights of Columbus interview, Betty alluded to the discovery of the neutrino to exemplify Clyde's faith. His patron saint was Saint Anthony, she said. As described on a Catholic information website, Saint Anthony was the patron saint of lost things. The description included a little jingle: "Saint Anthony, please look around; something is lost and must be found."[9] Betty said with a smile, "We figured that Saint Anthony helped him and Fred find the neutrino."

---

[7] Clyde Cowan, Catholic Scientist https://www.youtube.com/watch?v=45RMxpdEp6U.

[8] George Cowan interview, 2019.

[9] Catholic Straight Answers, http://catholicstraightanswers.com/who-is-st-anthony-and-why-is-he-the-patron-saint-of-lost-things/.

In 1967, the *Bulletin* of Catholic University published an article headlined "Threads of Thought," by Clyde Cowan. Nowhere in the article do the words "religion," "church," or "Catholic" appear. Yet the text appears to connect Cowan's understanding of science to his religious views.

A prominent physicist, the co-discoverer of the neutrino, invites the reader to spin with him some

THREADS OF THOUGHT

by Clyde L. Cowan, Ph.D.
Professor of Physics, Graduate School of Arts and Sciences

Bulletin, The Catholic University of America, vol. 1, no. 4, May 1967.

> It is the purpose of this brief essay to suggest a fresh approach to the problem of truth which, rather than calling upon mathematics or mathematical reasoning as guides, contemplates some concepts which lie at the base of all mathematics. Just as a piece of ground is not, *per se*, the game of baseball but furnishes the irreducible basis upon which a baseball diamond may be defined, rules constructed and a game begun, so there are concepts which are much broader than the mathematics, *per se*, and upon which they are built.
>
> We may approach the idea by considering the totality of the physical order — that universal domain commonly named Nature.... In the universal domain we include all phenomena as elements as well as all "ponderable matter," energy in all its forms, momentum in its every aspect. The human intellect is, of course, a set of phenomena included in the universal domain.
>
> We now reflect upon two major divisions of the universal domain, taking one of these as the totality of the human intellect, and the other as *all else* in the universal domain.

Cowan wrote that the two divisions comprised the "Totality of Nature" and that all of Nature exclusive of the intellect is knowable to that intellect. He used the term "mappings" to describe connections between knowable phenomena.

> In contemplating the mappings, we gaze into an infinity which is, itself, infinitely complex. For, certainly, some mappings resemble others, and correspondences exist within the set of all mappings which are themselves mappings.... The usual term for this state of affairs is "symmetry," and now we have gone full circle. For every symmetry known to the physicist, there is definable a natural law known to him as a "law of conservation" of some "invariant quantity." The physicist is among the first to assert, however, that the class of mappings with which he deals is but a small fragment of the vast set of correspondences which Nature presents to the intellect of man.[10]

Clyde Cowan's grandson, James Riordon, holds a bachelor's degree in physics and a master's in engineering. He became a science journalist and later headed public relations for the American Physical Society, a post he held for 15 years, until 2018. Born in 1965, James remembers his grandfather as a pleasant and gentle man. "He was very fond of children and he would take us to museums and talk about science with us." James said that his grandfather had an enormous influence on his interest in science and pursuit of a physics-related career.[11]

I asked James what he thought about the article that Clyde had written for the *Bulletin*. While proclaiming his own atheism, James had great respect for his grandfather's manner of thought. "I believe the article was consistent with his deep sense of spirituality," he said.

Members of the Cowan family are divided about whether Clyde's contribution to the discovery of the neutrino has been adequately recognized. But they are united about one conjecture. Like others in the physics world, they are confident that had Clyde been alive when the 1995 Nobel Prize was awarded, he would have been a co-recipient. Fred Reines

---

[10] Clyde Cowan, PhD, "A Prominent Physicist, the Co-Discoverer of the Neutrino, Invites the Reader to Spin with Him Some THREADS 0F THOUGHT," *Bulletin*, The Catholic University of America, May 1967, vol. 1, no. 4.

[11] James Riordon interview, 2019.

repeatedly spoke of his admiration for Clyde and his indispensable role in their joint discovery.

After Cowan's death in 1974, Fred recorded his feelings on a page that he titled "Letter to a Departed Friend." The undated letter was, of course, not intended to be mailed, and it remains in the Reines Collection at UC Irvine. Fred revealed a touching sense of loss that preceded Clyde's death. It had begun nearly two decades earlier, when their scientific collaboration ended.

> Dear Clyde,
>
> You should have stuck around. I could never figure out why you left — things were going so well, there was so much to do, so many questions to ask, so many answers to find. We were such a good team, the two of us — living for a few brief years inside each other's heads, sharing our sense of wonder at the unfathomed universe, looking at the uncharted seas with a childlike faith in the knowability of it all. We came from different poles to our intellectual partnership, our mutual assault on the unknown, our passionate pursuit crowned by that priceless reward — to find a new thing.
>
> It was lonely when you left, companionship makes the journey so much richer.... No one can understand better than you the excitement, the joy of a new thing, of standing on "a peak in Darien"* and seeing a new thing. But no, you went away and left me to fend for myself. To whom can I turn to share my dreams now that you are gone? You really should have stuck around.
>
> Your friend,
> Fred

*The phrase, "a peak in Darien," is from "On First Looking into Chapman's Homer," a poem by John Keats. It refers to the emotional impact on the Spanish explorer Cortez upon first seeing the beautiful Pacific Ocean from a mountaintop in the Darien region of Panama.

> Like stout Cortez when with eagle eyes
> He star'd at the Pacific — and all his men
> Look'd at each other with a wild surmise —
> Silent, upon a peak in Darien

In 1959, two years after Clyde left Los Alamos to pursue the next stage of his career, Fred departed as well. He would become chairman of the department of physics at Case Institute in Cleveland.

Part IV

Moving

Chapter 9

Moving to Case Institute

Case Institute, originally named the Case School of Applied Science, was founded in 1880. Seven years after it was established, physics professor Albert Michelson joined in a famous experiment at Case with Edward Morley, a chemist at the neighboring Western Reserve College. (The two schools merged in 1967 to become Case Western Reserve University.)

The Michelson–Morley experiment sought to detect changes in the speed of light as it traveled in different directions. Light was then thought to be propagated through an invisible medium called the ether. If a laboratory (attached to Earth) moved through the ether, the speed of light relative to the laboratory should vary depending on the direction of Earth's motion. The two scientists designed a device to compare the speeds of light beams simultaneously streaming at right angles to each other. To their surprise the speed of light in the lab appeared to be the same in all directions, unaffected by the motion of the Earth through the supposed ether.

The experiment showed unexpectedly that the presumed ether did not exist — or, if it did, was not detectable. The finding was material to Albert Einstein's 1905 formulation of special relativity: The speed of light is constant, no matter the speed of a traveling observer. The discovery also was the basis of the 1907 Nobel Prize in physics. When Michelson received the prize, he became America's first Nobel laureate in science. The experiment remains a source of pride at Case Western Reserve.

Fred's academic appointment created a new focus in the Case physics department. Some in the physics world had questioned whether Fred's reported findings of the neutrino at Savannah River were sufficiently persuasive, but most viewed the evidence as solid. Besides administrative and teaching duties, Fred had plans for further neutrino-related investigations. He assembled a team of faculty members and graduate students interested in neutrino research. The Neutrino Group at Case eventually grew to a dozen members. Among the first to join was William Kropp. In 1953, as a second-year undergraduate student at DePaul University in Chicago, he read in *Physics Today* about Reines' work. Kropp was riveted by the article and thought, "I'd like to meet this guy. But later, when I graduated, he wasn't around."

Bill Kropp later enrolled in the graduate physics program at the University of Maryland. Toward the end of his first year he learned that Fred Reines had joined the faculty at Case. It was 1959 and Bill decided to transfer schools.

> So I quit Maryland and went to Case Institute with the hope that Professor Reines would take me on. I went to his office and we talked for an hour or so. He told me I should go downstairs and find a place for myself in his laboratory. I was overwhelmed — it was great for me. This guy that I read about years earlier was willing to have me as a student. I worked with him ever since, until he died.[1]

Kropp shared this story with me in 2017. At 81, he was an emeritus researcher at the University of California, Irvine, where he remains an expert in neutrino physics. When I initially tried to reach him, his daughter Mary Ann, an attorney, answered the phone. I said that I hoped to talk to William Kropp about Fred Reines. She responded with excitement. He was unavailable then, but she was sure he'd want to speak with me. "My father loved Fred," she said.

Subsequently, during hours of conversation by phone, and in his home and office, I became accustomed to his soft voice and elfish grin. Kropp said that he regarded Fred as his scientific father. In talking to others who knew Fred well, I frequently heard similar expressions of veneration.

The neutrinos that Reines and Cowan had detected in 1956 were produced by a nuclear reactor. Their scientific positions at the Los Alamos

---

[1] Kropp interviews, 2017–2019.

National Laboratory enabled them to use the government-run reactors at Hanford and Savannah River. Although now at Case, Reines' status allowed him and his group to continue measuring neutrino properties at the Savannah River Plant. Fred had other ideas as well. He would look for neutrinos generated in the atmosphere and by the sun and other distant sources.

Identifying these "natural-source" neutrinos posed more challenges than detecting those from a human-made reactor. To find extraterrestrial neutrinos, he would need to overcome two key obstacles. The first was to block the confounding effects of unwanted cosmic radiation and other background activity; the second was to develop an apparatus that could detect neutrinos from sources more distant than a nearby reactor.

His new ambition was facilitated by a happy coincidence. A year before Fred's arrival at Case, the Morton Salt Company had opened a mine 30 miles northeast of Cleveland. The mine entry in Fairport Harbor reached half a mile down and it then extended horizontally beneath Lake Erie. Over time, the horizontal tunnel grew larger as explosives were used to loosen large chunks from the walls of salt. The chunks were crushed into smaller pieces, extracted from the mine, and sold commercially to melt road ice.

A 2016 article in *Wired* magazine described the mine's horizontal tunnel as totally dark except for the workers' headlamps. Three miles long, the tunnel is commonly traversed in a golf cart. The pathway eventually reaches "the main lode, an enormous space 40 feet wide and 20 feet high, with corridors stretching into darkness." A visitor "felt as if he'd stepped into another world, a hot, dry, odorless world where everything is white."[2]

The description would have been apt as well when the mine opened in 1958. Soon after Fred's arrival at Case, he entered the mine with some graduate students. They brought electronic equipment to measure the level of cosmic ray activity and the very low level of radioactivity generated by the salt. Fred was pleased with the result. Fred's colleagues later wrote:

---

[2] Laura Malonee, "Venture into a Surreal Salt Mine 2,000 feet below Lake Erie," *Wired*, May 3, 2016, https://www.wired.com/2016/05/venture-surreal-salt-mine-2000-feet-lake-erie/.

> He was satisfied that he had found a new experimental home. With a visit to Morton Salt headquarters in Chicago, he cultivated a working relationship with the owners of the company, an association that endured into the 1990s.[3]

The Morton company permitted Reines and his Neutrino Group to set up a laboratory in the mine. Investigations there ranged from developing a better understanding of beta decay and cosmic rays to the development of detection techniques, especially for neutrinos. Cherenkov detectors, instruments that display flashes of light induced by various particles, were refined for use in neutrino research. Articles that Fred authored in 1961 and 1962 suggested the range of his work in the mine. Their titles included "Scintillation Counters and Luminescent Chambers," "Neutrino Reactions," "A Large Cherenkov Detector," and "Experimental Test of the Conservation of Nucleons."

As Fred was searching for natural neutrinos, a group based at Columbia University was conducting a different kind of neutrino experiment. Leon Lederman, Melvin Schwartz, and Jack Steinberger were using a new particle accelerator at Brookhaven National Laboratory, 60 miles from New York City on Long Island. The machine, a high energy synchrotron, accelerated protons in an enclosed loop 600 feet in diameter. The particles were reined into their circular paths by powerful magnets. Before construction of the new synchrotron, the highest accelerator-driven energies could reach millions or a few billion electron volts. These were deemed "low energy." In the new synchrotron, the energy of protons could exceed 33 billion electron volts.

Through mid-20th century, more than one hundred subatomic particles had been identified. Many of them, such as pions, were produced by particle collisions in accelerators. The mean life of a pion is extraordinarily fleeting: 26 billionths of a second. Yet the pion was an important part of the Lederman team's experiment.

The experiment might answer a tantalizing question. Reines and Cowan's reactor-generated neutrinos were associated with electrons. Like electrons, muons are negatively charged, though a muon is 200 times

---

[3] *Ibid.*

heavier than an electron. (Like electrons, muons also have positively charged counterparts–antiparticles.) Would muon-associated neutrinos, if producible, be different from electron neutrinos?

Melvin Schwartz proposed the idea that a beam of neutrinos could be identified by an odd method of filtration: using a huge bulk of steel. The new Brookhaven accelerator seemed suited to provide an answer. The Lederman–Schwartz–Steinberger team was able to acquire 5,000 tons of old battleship plates. They lined up the plates to create a 30-foot thick steel wall. The synchrotron's high-energy protons could produce a shower of pions that would travel toward the steel wall 70 feet away.

During the brief trip, pions decayed into muons, which in turn decayed into electrons and neutrinos. Everything except the neutrinos was blocked by the wall. Only the neutrinos sailed through the entire thickness and then through a detector on the other side. As described on a Brookhaven website, "There, the impact of neutrinos ... produced muon spark trails that could be detected and photographed — proving the existence of muon-neutrinos."[4]

In an article titled "The Two-Neutrino Experiment," Lederman described the team's successful effort. The conclusion, he wrote, "is quite clear."

> There are two kinds of neutrino. Those produced by the decay of pions in our experiment are of the muon type and cannot produce electrons by interacting with neutrons. To produce electrons by this reaction one would need neutrinos of the electron type.[5]

The experiment was a breakthrough in neutrino studies, though it also raised new questions that would take more experiments and more years to resolve. Could there be additional forms of neutrinos beyond the two that were now identified? Were the two forms irreversibly different from each other or might they somehow transform from one type to the other? Does a neutrino have mass?

---

[4] Brookhaven and the Nobel Prize, https://www.bnl.gov/bnlweb/history/nobel/1988.php.

[5] Leon M. Lederman, "The Two-Neutrino Experiment," *Scientific American* March 1963: 60–70.

By 1962, when the two-neutrino finding was reported, Fred's Neutrino Group at the Case Institute had come to an unwelcome realization. Their search for natural atmospheric neutrinos would not be satisfied in their salt-mine lab.

True, at a depth of nearly half-a-mile, cosmic penetrations underground were vastly diminished. Compared with cosmic-source radiation on the Earth's surface, levels in the salt mine were lower by a factor of 40,000. Still, the holy grail eluded them. Efforts to detect atmospheric neutrinos were unsuccessful. Cosmic generated background events, though very much reduced, were still too plentiful to permit detection of distinctive neutrino trails. They determined that a deeper location would be needed to block more background interference.

Fred's search for alternative sites led him to consider locations around the world. He was especially interested in a report that cosmic ray intensity appeared to be negligible in a gold mine in the Kolar Gold Fields of India. At one-and-a-half miles beneath the Earth's surface it was much deeper than the salt mine near Case. The Kolar mine research team included scientists from the United Kingdom and Japan as well as India. Fred and Gus Hruschka, his engineer colleague, visited the mine early in 1963. They met with the Kolar scientists but were unable to come to an agreement to advance Fred's project.[6]

Then, in the summer of 1963, Fred identified perhaps the most promising location on the planet: the East Rand gold mine in South Africa. It was among the deepest of all mines. Fred contacted Professor Jacques Pierre Friederich Sellschop, director of nuclear physics research at the University of Witwatersrand in Johannesburg. "Friedel" Sellschop was enthusiastic about the possibility of participating in Fred's experiment.

Founded at the end of the 19th century, the university was closely associated with the mining industry. Its academic programs included numerous courses on mine engineering, mine design, and mineral resource management. The industry helped fund university programs that

[6] William R. Kropp and Marshall F. Crouch, "Some Reminiscences of the CWI [Case-Witwatersrand-Irvine] Atmospheric Neutrino Experiment," in W. Kropp, M. Moe, L. Price, J. Schultz, and H. Sobel, eds., *Neutrinos and Other Matters: Selected Works of Frederick Reines*. Singapore: World Scientific, 1990: 266–268.

in turn produced graduates who could help sustain and enhance mine activity.

The East Rand gold mine was in Boksburg, 15 miles from Johannesburg. As in other deep mines at the time, conditions in the mine included sweltering heat, leaks of poison gas, and explosive rock spurts. But with easy access to the university, support from the mine's proprietors, and an acceptable exploratory visit, Fred was eager to start his project there. Still, neither he nor other members of his Neutrino Group had anticipated some of the difficulties they were about to encounter. First, they had to deal with false allegations that their experiments involved nuclear weapons. Then, at Case, they would need to design and construct a detector larger than any previously built. Finally, they would have to deliver to the mine the detector, 144 photomultiplier tubes, tons of scintillator liquid, 5 miles of cable, and an array of other electronic gear.

Fred's detector had three main elements: scintillation liquid, tanks that held the liquid, and photomultipliers attached along the edges of the tanks. Neutrinos continuously pass through the Earth though on very rare occasions one will interact with another particle. If an interaction occurred during passage through the detector, the scintillator would display a faint configuration of light. Photomultipliers would enhance the light flash and transform it into an electronic signal. The signal could then be recorded and viewed on a monitor screen.

The Case scientists and technicians had been tasked with designing a huge detector that could fit into the mine. Size very much mattered. The larger the detector the greater the number of neutrinos that pass through it, increasing the chance that a rare neutrino interaction could be recorded.

The plan called for nearly 6000 gallons (22 tons) of scintillation liquid. No single tank containing that volume could fit through the mine's passageways. So the Case team built multiple plexiglass tanks, each about 18 feet long, two feet high, and six inches deep. More than 50 such containment tanks would be needed.

Bill Kropp reminisced that moving the equipment to South Africa was "complex" and "tangled." The tanks were packaged in wooden crates. In the course of shipment by air, sea, and land some crates were lost, stolen, misdirected, or held back because of maritime strikes. But these difficulties, he said, "paled in comparison to the problems associated with transporting the equipment through the labyrinth of the mine."[7]

Kropp described the effort:

> The crates were too tall to fit inside the elevator cage, so one at a time each was hung from the cage bottom, dangling below. The elevator descended very slowly to level 50, nearly a mile down. Then the crate was carried by hand a half-mile to the first of the inclined rail tracks. The crate was fastened to the top of a cart and brought down to the next level. There it was loaded onto another cart for another inclined ride. Loading on-and-off to get the equipment down the mine was quite a logistical nightmare.[8]

Shelving had been built in the mine laboratory to hold the plexiglass tanks stacked three high. Each unit, comprising three tanks, was six feet tall. The stacked units were placed end-to-end along the opposing walls of the dimly lit tunnel. The two parallel rows faced each other as if stationary mini-boxcars. Between the rows lay rail tracks on which a conveyance cart could roll.

The segmented detector took up most of the laboratory's space. The tanks would later be filled with scintillator liquid composed of clear mineral oil and chemical additives. When completed, the entire array would become the world's largest particle detector.

Subsequent to the Reines–Cowan experiment in 1956, new understandings about neutrinos had materialized. Neutrinos and antineutrinos were also associated with the decay of other unstable particles. Thus a pion instantaneously would decay to form a muon and a neutrino, as demonstrated in the Lederman–Schwartz–Steinberger experiment. Further, as Fred observed, "the muon in turn decays into an electron and two more neutrinos." The muon-related reaction would become essential in the search about to be undertaken in the East Rand mine.[9]

A plywood structure, the hut, had been constructed near the detector. It contained cables, vacuum tubes, and other electronic paraphernalia. Monitoring screens were installed to display and record scintillator

---

[7] W. Kropp, *et al.*, *Neutrinos and Other Matters*: 267, 318.

[8] Kropp interviews, 2017–2019.

[9] Frederick Reines and J.P.F. Sellschop, "Neutrinos from the Atmosphere and Beyond," *Scientific American*, February 1966: 41–43.

activity, and air conditioners in the hut reduced the temperature to a more bearable range inside the hut.

The group suffered a setback soon after the detector was in place when rock bursts crushed two of the plexiglass detector tanks, which had to be replaced. At another time the hut collapsed from the effects of a nearby dynamite explosion. Fred in Cleveland, and the crew in South Africa, were in continuous contact via teletype and telephone. Even after the hut was destroyed, they remained undaunted, as Kropp recounts:

> We had to start all over. We had built the hut ourselves and now had to do it again. After the hut was rebuilt we started installing the electronics. Everything was made from scratch. Made in-house. A lot of today's electronics didn't exist yet. Transistors were still not in general use. We used vacuum tubes.[10]

Despite the challenges, by the end of 1964 the apparatus was operational. The multi-unit detector could display not only the light pulse prompted by a particle, but the angle and direction of its passage through the apparatus. The Case-Witwatersrand research team now awaited the predicted interactions that would signify the presence of a muon neutrino.

Since muons carry an electric charge, their signals could be easily identified. Distinct from the plethora of muons streaming vertically, the detector was designed to identify muons "produced by neutrino interactions in the rock" that surrounded the laboratory.

During every second, trillions of particles presumably were passing through the detector. At first none exhibited the necessary directional and angular imagery to confirm an interaction. Finally, after more than 3 months of observation, there was a moment of euphoria. As described by Reines and Sellschop:

> Then on February 23, 1965, the detectors recorded a muon that had traveled in a horizontal direction — the first "natural" high-energy neutrino had been observed![11]

---

[10] Kropp interviews, 2017–2019.

[11] Frederick Reines and J.P.F. Sellschop, "Neutrinos from the Atmosphere and Beyond," *Scientific American*, February 1966: 41–43.

The exclamation mark at the end of the sentence reflected the excitement of the moment. The observation of just one particle out of trillions upon trillions made the difference. The research team immediately notified Fred, who was in the United States. He was jubilant. His project had led to the discovery of a new "thing." If not as momentous as his and Cowan's initial neutrino detection, the sense of accomplishment was no less thrilling. During the subsequent 11 months, nine more high-energy neutrinos were identified by the East Rand detector.

Around the same time, scientists at the Kolar gold mine in India reported a similar detection. The two independently conducted studies reinforced the validity of the finding. The reported detection of natural neutrinos had been confirmed.

Neutrino Detection Laboratory at Bottom of South African East Rand Gold Mine.

Source: Frederick Reines and JPF Sellschop, "Neutrinos from the Atmosphere and Beyond," *Scientific American*, February 1966.

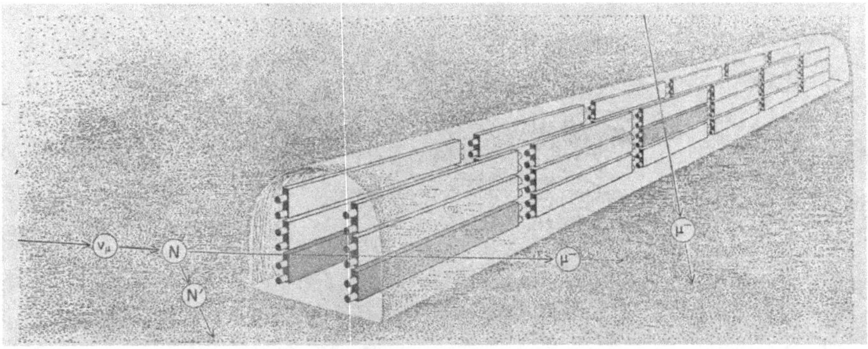

NEUTRINO-DETECTION ARRAY currently in use at the authors' South Africansite.... Two muons [μ^-] are shown being detected. Entering from the left and passing through a detector in each row is a daughter muon produced by the interaction of an atmospheric muon-type neutrino [V_μ] with the surrounding rock.[12]

[12]F. Reines, M. F. Crouch, G.M. Frye, T.L. Jenkins, R.M. Woods, W.R. Kropp, H.S. Gurr, "The South African Deep Mine Neutrino Experiments," Technical Report [unpublished], Research Program in Neutrino Physics, Cosmic Rays and Elementary Particles, Atomic Energy Commission, September 1966.

Chapter 10

New Dean, New University

When the University of California at Irvine opened in the fall of 1965, Kenneth Ford, the quick-witted chair of the new Physics Department, displayed a broad smile and a head of dark hair. In 2018, at age 91, his hair was white, but his intelligence and smile were undiminished. He had recently published his tenth book, *Building the H Bomb*, drawn from his experience nearly 70 years earlier.

In 1950, Ford was a graduate student at Princeton. His faculty adviser, John Wheeler, would be going to the Los Alamos Laboratory for a year, to participate in the development of the hydrogen bomb. Ken accepted an invitation to join him. "I thought of the United States as a moral nation," he recalled, "and the Soviet Union as a nation that could not be trusted." He also relished the opportunity to remain close to Wheeler, "whom I so much admired."[1]

During his year in Los Alamos, Ford, like other young researchers, worked on computations toward building a hydrogen bomb. Their efforts ranged from evaluating possible modes of ignition, to deployment of fission and fusion components. He said the work was interesting and he enjoyed associating with the many notable figures at the laboratory. Stanislaw Ulam was still employed there, and the retinue of other icons who periodically visited included Enrico Fermi, Hans Bethe, Edward Teller, and John von Neumann.[2]

---

[1] Kenneth W. Ford, *Building the H Bomb, A Personal History*. Singapore: World Scientific, 2015: 34–35.

[2] *Ibid*: 105; Ford interviews, 2017–2019.

Fred Reines, like Ulam, still full time at the Lab, had not yet begun his project on the neutrino and Ken Ford knew him only casually. Ken remembered a large room where people would gather for coffee, conversation, and lectures. For reasons that he could not fathom, it was called the *Reinesraum*, German for "Reines Room." Perhaps, Ken thought, the name was related to Fred's large presence. "He had a booming voice, both speaking and singing, and his general enthusiasm was infectious," Ken observed.

> He [Fred] would drop in at the *Reinesraum* to chat and trade ideas. For most people, it was impossible not to like Fred. My mentor, John Wheeler, for instance, liked Fred because he "thought big."[3]

After a year in Los Alamos, Ken returned to Princeton to do further work on fusion energy and to complete his PhD requirements. He got to know Fred better when he returned to Los Alamos during the summer of 1953 and again in 1957. The latter visit occurred a year after Fred and Clyde Cowan had detected the neutrino. Fred "was in even higher spirits than usual," Ken recalled. He also witnessed Fred's expansive talent in Gilbert and Sullivan operettas in which Fred had leading roles.

Seven years later, in 1964, Ken left a faculty position at Brandeis University to become chairman of UC Irvine's new Physics Department. The following year he also became acting dean of the new Division of Physical Sciences. The deanship involved oversight of the division's three new departments: chemistry, mathematics, and physics (later expanded to physics & astronomy). The added responsibilities interfered with Ken's teaching, research, and other administrative tasks. He was pleased to head the search committee for a new dean.

> The first person I thought of for the job was Fred Reines. The committee agreed, and, after Fred visited us, the Vice Chancellor for Academic Affairs (then Jack Peltason) also agreed, and, best of all, Fred agreed… . Unlike me, Fred wasn't afraid that the deanship would put an end to his physics. He made several requests that were honored. On the administrative

[3] Kenneth Ford interviews and email, 2017–2019.

side, he asked permission to hire a high-level administrator to handle most of the daily duties of the dean.[4]

Coincidentally, the house next to Ken's had come on the market and Fred and Sylvia bought it. The two men were not only professional colleagues, but the Ford and Reines families also became good friends. Ken nodded in agreement when, 50 years later, his wife, Joanne, told me that "Sylvia was a lovely person and Fred was always cordial and very gracious."[5]

During a party one evening, Fred suddenly dropped to one knee in front of Joanne. She was startled as he began to sing an aria in a language she didn't understand. Ken and the other bemused guests were also uncertain about the language, and whether Fred was acting out a proposal for marriage or perhaps begging for forgiveness. The moment remains vivid in Joanne's memory. "Fred was bigger than life," she said, "and he had a wonderful voice."

Fred's move to Irvine in 1966 added a new base for neutrino and other particle research. Until then the experiment in the Rand mine had been a Case Institute project that was supplemented by Witwatersrand. Now work there would become a Case-Wits-Irvine collaboration. Fred's protégé, William Kropp, had moved with his family from Cleveland in mid-1964 to head the group in the mine laboratory. Soon after, Henry Gurr, his wife, and 6-week-old son, arrived in South Africa as well.

Gurr, who grew up on an Ohio farm, had a 1959 undergraduate degree in engineering science from Case. He began graduate school there in physics just before Fred arrived. In 2018, speaking through a white beard that reached his chest, Gurr, now 81, recounted his early impressions of Fred:

> Everybody at Case was so amazed that the school had gotten Fred Reines. He had found the neutrino with Cowan. He had accomplished the impossible. I went to a seminar he gave and still thought what he had done was incredible.[6]

---

[4] *Ibid*.

[5] Joanne Ford interview, 2017.

[6] Henry Gurr interview, 2018.

Although not especially interested in the neutrino, Gurr decided to ask Fred to be his academic adviser. He was aiming for a master's degree. To his own surprise that goal changed during his interview in Fred's office. Fred said he would take on graduate students only if they were working toward a PhD. "Are you going for your PhD?" "Well, uh, yes," Gurr replied.

> And that was the beginning of one heck of a ride. I can tell you, Fred Reines was one heck of a leader.... You wanted to get done for him the things he wanted you to do. You would do your best and you didn't want to let him down.

More than a half century later, Gurr still seemed in wonder about Fred's leadership.

> He didn't crack a whip and say "Get busy"…. He'd never say, "Well, you made a mistake".... He might suggest another way to do something. If you really did something [disagreeable] he wouldn't say anything. He'd just hesitate and close his lips tight together. You knew that you were off track and that you'd better straighten it out.

Gurr's PhD obligations included helping to set up the detector and electronic wiring in the South African mine laboratory. His dissertation validated the finding that protons in an atomic nucleus are perhaps stable, or at least highly resistant to disintegration. Using data from the detector, he showed that the proton's "mean life (if it has one) is vastly greater than the age of the universe." (As of this writing, the half-life of the proton has still not been firmly established, but is believed to be more than 10^{34} years, hundreds of times longer than Gurr's upper limit.) Gurr returned to Cleveland in 1966 to receive his PhD at Case. With Fred's encouragement he moved to Irvine.

Fred was again overseeing projects at the Savannah River reactor site near Aiken, South Carolina. Gurr joined an experiment there on neutrino scattering. The aim was to establish that neutrinos coming from a nuclear reactor would collide with electrons at a rate predicted by a theory of Richard Feynman's. Eventually, after years of effort, the team succeeded. But in the course of those years, Gurr was struck by a particular incident.

At one point, when the reactor was on, the detector seemed to pick up a huge number of neutron signals. As a result, "we were supposedly seeing a thousand times more neutrinos than the predicted number." Gurr thought the observation may have been caused by background sources that mimicked neutron signals. But Fred would not countenance any doubts. As Gurr recalled:

> Reines was saying, "Oh my God, this is a Nobel Prize. This is super, super." He's telling everybody how happy he is. I kept telling him, "uh, uh." We were seeing little blips on the oscilloscope, but we don't know what they are. Finally in retrospect we realized they were consequences of gamma rays. Reines kept saying it was neutrinos bouncing around, neutrino scattering. He was so excited that he seemed to abandon caution.[7]

Other members of the Neutrino Group also tried to convince Fred that he could be wrong. Still, he resisted. Gurr said that during the next few weeks "he calmed down and realized he had been too overjoyed."

While working at Savannah River, Gurr taught some courses at the University of South Carolina in Aiken. He became increasingly interested in methods of teaching. After completing his neutrino experiment in 1975, he continued to divide time between Irvine and Aiken. In 1980, he became a full-time professor at the University of South Carolina. His academic focus had shifted from physics research to teaching and exploring the basis of creativity.

The courses he taught on physics and mathematics included readings on the evolution of consciousness, as well as on wind instability and sound formation. Gurr's unusual though captivating subject matter prompted the school to honor him with the title of distinguished professor. Even after veering from nuclear research he maintained the habits of inquiry acquired under Fred's guidance. "Fred was a tough taskmaster, but he was such a good leader that you didn't realize it." Gurr reflected again on the profound influence that Fred on him: "He didn't criticize. He was just immensely interested in what you were doing."[8]

Another young member of the Neutrino Group initially at Case and later at Irvine was Henry Sobel. Born in Brooklyn, Hank attended Rensselaer Polytechnic Institute in upstate New York. At RPI he read about

---

[7] *Ibid.*

[8] *Ibid.*

the discovery of the neutrino and wrote to Fred Reines with some questions. Fred answered, and enclosed an application to graduate school. Sobel entered Case in 1962. In 1966, as Henry Gurr had done, he moved to South Africa to complete his PhD thesis. Besides studying atmospheric neutrinos, Sobel also began an experiment to find neutrinos coming from the sun.

Fred had been thinking about ways to detect solar neutrinos as well as those created in the Earth's atmosphere. The Reines Collection at UC Irvine includes pages of his hand-written equations, cross-outs, and messy notations. One page, headed "Solar Neutrinos," is dated September 3–4, 1964. Next to flocks of numbers and symbols are apparent reminders: "Use crude assumed responses vs. distance … to get general idea." "Need to know angle rate vs. energy up to 6 MeV [million electron-volts] for one photomultiplier tube." At the bottom of the page in large block lettering is an admonition to himself: "MUST STUDY THIS FURTHER." Whatever the charts and numbers meant to him, Fred apparently was pondering how the detector could be used to identify solar neutrinos. He wondered whether inverse beta decay or a different mode of detection would be suitable.

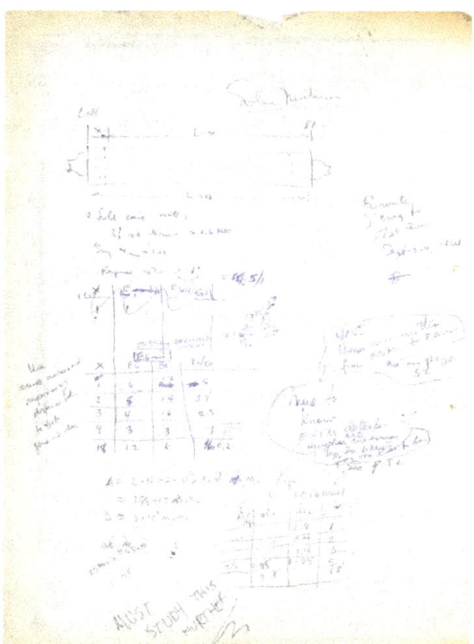

Reines' notes on possible approaches to solar neutrino detection, 1964. Reines Collection.

In considering various chemicals for use in a detector's scintillation liquid, Fred had noted: "The neutrino capture cross-section is considerably higher for lithium than for boron." The event rate (frequency of interactions) per ton of lithium-producing electrons was estimated at 130 per year and the estimated rate for boron-producing electrons was 24 per year. Trillions of neutrinos pass through the detector every second. Thus the event rate, using either element, appears spectacularly minuscule.[9]

The processes in the core of the sun presumably apply to other stars as well, so confirmation of the existence of solar neutrinos could be a leap toward understanding the cosmos. Hydrogen, the most abundant element in the sun (and in the universe) also has the simplest atomic structure. It consists of one electron outside of a nucleus composed of one proton (or sometimes one proton and one neutron — the nucleus of so-called heavy hydrogen). Immense pressure and temperature (27 million degrees Fahrenheit) in the sun's core, force hydrogen nuclei to fuse. The fusion creates helium nuclei, consisting of two protons and one or two neutrons. The process also generates massive energy in the form of light and heat.

Fred knew that a colleague, Raymond Davis, had independently been trying to detect solar neutrinos. Davis was exploring experimental procedures that had been suggested, though not carried out, by Bruno Pontecorvo in the 1940s. To demonstrate the presence of neutrinos the method would involve a neutrino collision with the nucleus of a chlorine atom. An interaction would transform the chlorine nucleus into an argon nucleus, which would then form an argon atom. (Chlorine and argon are next-door neighbors on the periodic table.) The argon isotope could be easily extracted from a detector filled with a liquid containing chlorine. Davis' liquid of choice was perchloroethylene, a common dry-cleaning solvent. Because neutrino interactions are so rare, a very large volume of the solvent would be necessary.

---

[9]Raymond Davis, "Fred Reines and Solar Neutrinos" in C.E. Lane and R.I. Steinberg, eds., *Discovery of the Neutrino*, Franklin Symposium Proceedings, April 30–May 1, 1992. Singapore: World Scientific, 1993: 8.

Davis conducted a pilot study using a 1000-gallon tankful of the solvent in a 2200-foot-deep Ohio limestone mine. He could not detect solar neutrino signals. He surmised that the tank was too small and the mine insufficiently deep to block background interference. In the early 1960s John Bahcall joined Davis in his search. Bahcall calculated that a few atoms of argon per week could be captured in a large-enough tank of cleaning fluid. In 1965, Davis and Bahcall completed placement of a 100,000 gallon detector nearly a mile underground. They constructed the holding tank deep in the Homestake gold mine in South Dakota. In 1968–1969 they detected neutrinos from the sun.

Raymond Davis (left) and John Bahcall at Detector in Homestake Gold Mine, South Dakota.
Photograph from J. N. Bahcall, "Solar Neutrinos," Proc. 2nd Int. Conference on High-Energy Physics and Nuclear Structure, G. Alexander, ed., (North Holland, Amsterdam), 1967.

The finding was staggering. Until then, visible information about the sun was limited to what could be seen on its surface. The sun's energy, including its heat and light, is generated in its core. To reach the sun's outer surface, the heat and light must travel through a "seething plasma" of tightly packed particles. The trip outward to the surface could take a thousand years or more.[10]

In contrast, energy in the form of neutrinos travels nearly at the speed of light and they are not blocked by any amount of mass. The neutrinos produced in the sun's core reach its surface in a few seconds. Thus light could take a thousand years to reach the sun's surface and another eight minutes to reach the Earth. But a neutrino could make the entire trip in little more than the eight minutes. "Our sole motivation," Bahcall wrote, "was to use neutrinos to enable us to see into the interior of a star and thus verify directly the hypothesis of nuclear energy generation in stars." In this they succeeded.[11]

There was a hitch, however. The number of solar neutrinos detected by Davis and Bahcall was only one-third the number predicted by Bahcall. The "solar neutrino problem" remained unresolved for 30 years. It was later found that the neutrinos emitted by the sun — electron neutrinos — could "oscillate." That is, they could change their identity into other kinds of neutrinos, called muon and tau neutrinos, during passage to the Earth. But the important fact was that Davis and Bahcall had confirmed that the sun emits neutrinos. Given the confirmation, Fred and his team suspended their efforts and Henry Sobel turned to other neutrino experiments in the Rand mine.

Sobel left South Africa in 1969 but continued to be a part of Fred's Neutrino Group, now based at Irvine. He spent much of his research time with Henry Gurr and other group members at the Savannah River site, where they were seeking to identify "neutrino-electron elastic scattering." Sobel describes the phenomenon, as did Gurr, as "a neutrino bouncing off an electron like a billiard ball collision." Such an event is very difficult to identify because it has a very small cross-section — it happens infrequently. Sounding relieved, Sobel noted, "In the 1970s we finally managed to design an experiment that was able to do it."

[10] "Ancient Sunlight," NASA, 2007, https://sunearthday.nasa.gov/2007/locations/ttt_sunlight.php.

[11] Bahcall, "How the Sun Shines"; Davis, "Fred Reines and Solar Neutrinos," in C.E Lane and R.I. Steinberg, eds., *Discovery of the Neutrino: Franklin Symposium Proceedings, Philadelphia, April 30–May 1, 1992*. Singapore: World Scientific, 1993.

In 2020, Sobel, now an emeritus professor at Irvine, remains at work on several neutrino-related projects. Among his positions, he is a representative in the United States of the giant neutrino detection laboratory in Japan, called Super-Kamiokande. He advocates for financial support for continuing American participation in Super-K experiments.

Unlike Kropp, Gurr, Sobel and other early members of the Neutrino Group, Michael Moe never worked in the South African mine. Yet his story produces a window into Fred Reines' extraordinary support for his students and colleagues.

After completing high school in Milwaukee, Wisconsin, Moe attended Stanford University where in 1959, he received a Bachelor of Science degree. He arrived at Case Institute to pursue graduate studies soon after Fred had become chairman of the physics department. The new chairman was reputed to be intensely devoted to physics and very demanding of his graduate students. He was known to phone a student in the middle of the night, when feeling the need to share a compelling idea. Moe disregarded advice by a fellow-student to find a less demanding faculty adviser.

> At my interview Dr. Reines looked at me and said, "You look like you could lift a lead brick." Turns out he wasn't kidding. I must have lifted, literally, 50 tons of bricks working for him.[12]

The bricks, each about 25 pounds, would be stacked around a detector to block unwanted background radiation. They created a wall of lead to shield against penetration by gamma rays emanating from the reactor.

> I guess I was a little naïve in my expectations of what it would be like to work for Dr. Reines. I thought the customary summer break from classes would be an opportunity for plans to travel west. He said, "No, that's not an option." I relented and abandoned my travel plans.

Moe received his PhD at Case in 1965. His dissertation was based on testing a law of conservation, an abiding interest of Fred's. Moe's experiment, urged by Fred, involved looking for electron decay. Since an electron is the lightest particle with an electric charge, presumably there

---

[12] Michael Moe interview, 2018.

is nothing that it can decay into. Finding evidence to the contrary would violate charge conservation. A fundamental precept of physics would have been challenged.

After trying different techniques, in the end he saw no electron decay. Still, his skillful efforts merited receiving his doctorate. Moe had worked largely in the salt mine laboratory near Cleveland where previous efforts to detect atmospheric neutrinos had been unsuccessful.

Following a postdoctoral year at the California Institute of Technology, in 1966, Moe rejoined Reines and the others from Case who had moved to UC Irvine. He was appointed assistant professor and spent time at the Savannah River plant helping with the neutrino scattering experiment. In the early 1970s, he read a preprint of an article sent to Fred by Dr. C. S. (Chien-Shiung) Wu. Madame Wu, as she was known, was at Columbia University, where she had previously worked on an aspect of the Manhattan Project. The diminutive and brilliant Chinese émigré had become an expert on beta decay and her article was about an extremely rare reaction called "double beta decay." The postulated phenomenon occurred when two neutrons simultaneously transform into two protons, two electrons (beta particles), and two neutrinos.

Madame Wu's article reported an inability to observe the reaction because of background interference from a ubiquitous radioactive isotope of bismuth. Moe recalls:

> I had some ideas about how to identify the bismuth as an imposter of the double beta decay signal, so I wrote a proposal. I was up for tenure and promotion to associate professor. So a decision was about to be made whether or not to offer me an associate professorship.

Moe submitted his proposal to the Irvine physics department in 1972. The department chair sent a copy to Madame Wu with a request for a letter of recommendation for an associate professorship. Moe's voice lowers:

> Madame Wu was not impressed with my idea. In her reply to the department chair she mentioned another contaminant, actinium, which I had not mentioned with my technique for suppressing background. She said my technique would fail because this other isotope would evade my method of rejection.

As a consequence, Moe was summoned to the office of the department chair. He was not to be promoted, which meant denial of tenure. "Hearing that was hard. I was deeply disappointed," he said after a long pause.

Nonetheless, Moe believed that by eliminating bismuth he might still see proof of double beta decay. Reines agreed and was determined to support Moe's research. Undeterred by the department's decision, Fred offered him a position as a research physicist. He would fund Moe from his own grant money.

> Fred supported me while I went ahead with the double beta decay research, so I owe him a lot because he stood by me for 15 years while I did that experiment. It took me that long to get the result.

Moe sought double beta decay in an isotope of selenium, an element he thought might show the reaction. The decay associated with selenium occurs so slowly that its half-life is scarcely conceivable — many times longer than the lifetime of the universe. Moe was finally able to detect the release of a neutrino from each of the two neutrons that are part of the process. The identifying signals were tracks of two beta particles that can be detected moving away from a decaying atom.

Given the extremely long half-life of the selenium isotope, a reaction among a sea of selenium atoms might be observed about once a week. Alan Hahn, a colleague of Moe's, described the isotope as apparently having "the longest half-life of any radioactive substance ever discovered."

The importance of Moe's discovery was underscored in 2013 when he received the Tom W. Bonner Prize in Nuclear Physics. The prize is awarded annually by the American Physical Society for a significant contribution to nuclear physics research. The citation stated that Moe's work had "transformed the field."[13]

Moe's eventual success in 1987 also bore an irony: Not only did he and his collaborators confirm the existence of double beta decay, "We

---

[13]Malcolm W. Browne, "Predicted Decay of Element Seen," *The New York Times*, September 4, 1987; Dennis Silverman, "Michael Moe Honored for the Discovery of Double Beta Decay at UC Irvine," *Energy Blog*, May 31, 2013, http://sites.uci.edu/energyobserver/2013/05/31/michael-moe-honored-for-the-discovery-of-double-beta-decay-at-uc-irvine/.

showed that actinium was not a significant background." Madame Wu's caveats were proved unwarranted.

In thinking back on his long association with Fred, Moe recalled some of Fred's seemingly idiosyncratic behaviors. For many years, Fred was always *Dr. Reines*. "There was no first-name basis. I was Mr. Moe and all the other students were Mr. as well." Fred's enthusiasm was contagious, yet students might cringe when he entered the laboratory. He would "test our electronics by yanking hard on the cables and by bouncing power supplies off the bench." His exuberance could be misplaced, as when he mistakenly embraced unconvincing results of the neutrino experiment described by Henry Gurr.

But overall, Moe viewed Fred as a man of vision and dreams. "He taught us to maintain the momentum of a project with impatient zeal, and to steamroller over all obstacles, especially those of a bureaucratic nature."[14]

Moe never became a professor. In retrospect, he considered the absence of the title to be fortunate. Undistracted by teaching and committee responsibilities he was able to devote full time to his experiment. "And Fred made that possible. It wasn't always obvious that the experiment would be successful, but he stood by me through thick and thin."

[14] Michael Moe, "Fred Reines as Teacher: Recollections of a Former Graduate Student," in W. Kropp, *et al.*, *Neutrinos and Other Matters*: 539–540.

Chapter 11

Transition Years at Irvine

At the outset of the 1970s, Fred Reines could view the previous decade with satisfaction. After detecting atmospheric neutrinos in the South African mine, his research continued to prosper both there and elsewhere. The transition from Case Institute to the University of California, Irvine, in 1966 was proving fruitful. As founding Dean of Physical Sciences he

Fred Reines and members of the Neutrino Group, 1969. Reines Collection.

navigated the growth of the new departments of chemistry, mathematics, and physics. His reputation as a neutrino expert attracted new faculty members and student enrollees to the physics department. A half-dozen of Fred's colleagues at Case transferred to Irvine or continued working with him from Case. Toward the end of the 1960s his Neutrino Group at Irvine had expanded to some 20 members.

Reines and the Neutrino Group on 83-Ton Electromagnet, UC Irvine, c. 1969. Reines Collection.

The new Irvine campus was growing during this time of social upheaval in America and beyond. Thousands demonstrated against government policies in cities across Europe. Dissenters filled the streets in Soviet-dominated states — Poland, Czechoslovakia, Yugoslavia — but also in democracies including the United Kingdom, France, and Italy. In the United States, continuing discrimination against black people had prompted violent reactions in several cities. At the same time, protests against the Vietnam War were growing. By the end of the decade dissatisfaction with social, political, and economic policies had fueled rebellion on numerous college campuses. Events in 1968 were particularly unsettling. During that year Martin Luther King, Jr. and Robert Kennedy were assassinated. Violence at the Democratic Party Convention in Chicago generated a sense of national instability.

The wave of demonstrations on American college campuses began in 1964 with the "free speech movement" at the University of California, Berkeley. By 1968, student-led protests had spread to colleges throughout the country, including the nascent Irvine campus. Dissent there had not become violent, as it had at some other leading schools, including Columbia, Cornell, the University of Wisconsin, and Berkeley. However, in January, when the campus newspaper, *The Anthill*, published an article about Reines' association with the University of Witwatersrand and his research in the nearby mine, Fred became distressed.

The article prompted several students and professors to question Fred's action. *The Anthill* published an "open letter to Dean Frederick Reines" by five philosophy professors.

> We are surprised to learn of the university's participation through your agency with a South African university. The article in *The Anthill* of January 23 gives few details but, if it is true then at best you have involved us and the AEC [Atomic Energy Commission] with the institutions of a country whose policy is explicitly and oppressively racist in the worst traditions of white colonialism. Our concern is not with scholarly communication, which we do not wish to censor. It is that our involvement may have political and economic features which go considerably beyond such communication.
>
> We recognize that certain advantages for scientific investigation may accrue from such a privileged location. We must ask you to recognize, however, that there are excesses in the name of science, and we must ask you to explain to us as colleagues who are your unwitting accomplices the extent to which you have committed us to support or acquiesce in the affairs of a state whose policies we find disgusting.[1]

[1] *The Anthill*, February 9, 1968.

Unlike African delegates at the UN in 1964, the Irvine critics did not make false allegations about the nature of the underground experiments. There were no claims about nuclear weapons testing, as four years earlier at the UN. The philosophers' objections were based solely on South Africa's apartheid policies. Fred's response was conciliatory. Still, he insisted that his scientific work there was important, open, moral, and had nothing to do with politics. His reply to the philosophers:

> Your open letter regarding the possible political and economic overtones of our cosmic ray and neutrino experiment being conducted in a deep mine in South Africa requests an answer.
>
> The experiment is being conducted there because there exists no other such location in the entire world in which to carry on these scientific cosmic ray investigations. Were we to prepare a comparable location in the United States it would cost the resources of the magnitude required to build a small university.
>
> You recognize in your letter the importance of scholarly communication and indeed this is a fitting description of our activities. The work being done is publishable in the open literature, free to all interested scientists throughout the world. As to the political aspect — we are all observers of a society beset by deep problems. In this role we are not party to interference pro or con in how these problems are to be resolved.
>
> Our presence in South Africa, necessary to do this scientific work, neither condones nor condemns the policies of the state, whatever our private views might be.
>
> I thank you for raising these questions indicating an awareness and deep concern for our moral stance in the world. It is proper that we all thoughtfully examine the consequences of our actions whether they be motivated by scientific curiosity or the desire for economic gain. It is my considered judgement that work in no way commits me, the university or you to the social structure of South Africa.

Whether the reply satisfied any of Fred's critics is uncertain. Substantively, it was a polite reaffirmation of his view of the primacy of science. In any case, funding for the East Rand mine laboratory was shrinking. Grants from the U.S. Atomic Energy Commission, which had been crucial to the project, were not being renewed. Research in the mine ended a few years after Henry Sobel left South Africa and returned to Irvine in 1970.

The challenge to Fred's South African mine project had been an unpleasant distraction. But 1968 was also a time of good fortune for his Neutrino Group. An engaging young physicist had arrived at Irvine. Born in wartime China in 1942, Herbert Chen had immigrated with his family to the United States in 1955 under a recently enacted refugee entry law. After high school in Massachusetts, he attained a bachelor's degree from the California Institute of Technology in 1964. Four years later he received a doctorate in physics from Princeton.

Chen proved to be an unusually creative addition to the Irvine physics department. In 2018, Professor Jonas Schultz, whose research team had collaborated on an experiment with Fred's Neutrino Group, remembered Chen as a dedicated scientist and teacher.

> It was obvious to everyone how outstanding he was. Fred welcomed that kind of initiative. He supported Herb initially out of his own funds until Herb could go out and get funding on his own.[2]

Chen had come to Irvine as a theorist, but after a couple of years he was developing ideas for neutrino experiments. In 1971, Fred co-authored a paper with Kenneth Lande that proposed the building of a powerful linear accelerator at the Los Alamos Meson Physics Facility (LAMPF). Construction was undertaken and completed in the late-1970s. Later, Chen temporarily moved to Los Alamos to lead a key experiment planned for the new machine. His experiment attempted to confirm a predicted, though yet unobserved, phenomenon — an "electron-neutrino scattering reaction" by using an accelerator-source of neutrinos.

A neutrino bumping into an electron would have the effect of a billiard ball collision. As a result, the energy of each colliding particle would be modified. The interaction had already been observed involving reactor-source antineutrinos by Fred Reines and members of his Neutrino

---

[2] Jonas Schultz interview, 2018.

Group.[3] At LAMPF Herbert Chen's group became the first to observe electron scattering by an accelerator-source of neutrinos.[4] Chen's other projects also proved successful and he had become recognized as a leader of the neutrino physics community at Los Alamos.[5]

In the 1970s a new conceptualization of particle physics, known as the Standard Model, provided insight into the fundamental structure of matter. As explained on the CERN website:

> Everything in the universe is found to be made from a few basic building blocks called fundamental particles, governed by four fundamental forces. Our best understanding of how these particles and three of the forces are related to each other is encapsulated in the Standard Model of particle physics.[6]

Among the fundamental particles are quarks, the building blocks of protons and neutrons. Other fundamental particles include the electrically charged electron, muon, and tau, and also the electrically neutral electron neutrino, muon neutrino, and tau neutrino.

As reported on the Fermilab website, the tau neutrino was not directly detected until 2000.[7] Actually, Harvard physicist Gary Feldman had found evidence of its existence some years earlier, when he was a graduate student in Martin Perl's group at Stanford.[8] But even before then, its existence had been presumed, according to the Standard Model. This presumption had been important in addressing a problem that had been frustrating particle physicists, regarding solar neutrinos.

---

[3] F. Reines, H.S. Gurr, and H.W. Sobel, *Physical. Review Letters.* 37(1976).

[4] R.C. Allen, *et al.*, "Study of Electron-Neutrino Electron Scattering at LAMPF," *Physical Review.* D47(1992).

[5] K. Lande, F. Reines, Neutrino Facility Proposal, LA-4842-MS, Los Alamos Scientific Laboratory of the University of California, Los Alamos, NM, https://permalink.lanl.gov/object/tr?what=info:lanl-repo/lareport/LA-04842-MS; Gerry Garvey, "A Brief History of Neutrino Experiments at LAMPF," *Los Alamos Science*, No. 25, 1997, https://permalink.lanl.gov/object/tr?what=info:lanl-repo/lareport/LA-UR-97-2534-06.

[6] The Standard Model, CERN, https://home.cern/science/physics/standard-model.

[7] Physicists Find First Direct Evidence for Tau Neutrino at Fermilab, July 20, 2000, https://news.fnal.gov/2000/07/physicists-find-first-direct-evidence-tau-neutrino-fermilab/.

[8] Gary Feldman interview, 2019.

In 1969, when Raymond Davis detected neutrinos coming from the sun, he found only one-third the quantity predicted by John Bahcall's theoretical calculations. Accounting for the "missing" neutrinos was known as the "solar neutrino problem." During the late 1970s, the problem had piqued Herbert Chen's interest. Chen's curiosity would later lead to the establishment of the Sudbury Neutrino Observatory in Canada. Dubbed SNO and described more fully in Chapter 16, the observatory was instrumental in solving the problem.

Chen's rising stature in the physics world was vigorously supported by Fred Reines. Herbert's daughter, Christine Chen, described their relationship as one of mentor and mentee. Now an astronomer at the Space Telescope Science Institute in Baltimore, Christine recalled that during the 1970s Fred helped create several opportunities for her father.

Herbert Chen, c. 1980. Reines Collection.

> My dad would have ideas and he would talk to Fred about them and would ask if Fred would support them. Fred would see people at a conference and say to them, "Here's my colleague, Herb Chen. He has this great idea. Herb, you should tell them about it."[9]

[9] Christine Chen interview, 2018.

Toward the end of the decade Fred received a request that had nothing to do with science. It began with a phone call from Howard Lenhoff, a distinguished biological scientist at UC Irvine. Lenhoff, an admired humanitarian, was a leader of the American Association for Ethiopian Jews. The black Jews of Ethiopia traced their religious forebears to ancient Israel. An impoverished minority, they had long been discriminated against by their Christian and Muslim neighbors. In the 1970s, the chief rabbis in Israel acknowledged their right as Jews to immigrate to Israel. The Ethiopian government would not permit them to leave, nor did the Israeli government seem particularly eager to receive them. In 1978, Lenhoff presented Israeli Prime Minister Menachem Begin with a petition calling for their rescue.

Three years earlier, Lenhoff came across a 1906 letter by prominent Jews including 44 leading rabbis in Europe, America, and Egypt. The letter, to the Ethiopian Jewish community, promised support for their religious education and for their aspiration to return to Israel. Among the signatories was the great rabbi from Russia, Isaac Jacob Reines. The name struck Lenhoff.

As Lenhoff later wrote, he realized that "on our campus there was a prominent physicist . . . named Frederick Reines. He and I had been active soliciting funds among Jewish faculty to assist Israel during the Yom Kippur War [in 1973]. I called Reines and he told me that he was a grand nephew of the Rabbi."

Lenhoff initiated a petition on behalf of the Ethiopian Jewish community, suggesting that the petition be linked to a commemoration of the 1906 letter. Fred agreed to be the first to sign. A photograph in the *Los Angeles Times* shows a smiling Fred Reines with pen in hand above the document.[10] When the petition was presented to Begin in 1978, it included the signatures of 50,000 American Jews. Lenhoff told Begin that the petitioners represented only a fraction of American Jews who supported the effort, and Begin was touched that so many American Jews cared.[11]

---

[10] Gordon Grant, "Ethiopian Jewish Group: Concern for 'Poorest of Poor' Alive," *Los Angeles Times*, August 17, 1975: OC3.

[11] Howard M. Lenhoff, *Black Jews, Jews, and Other Heroes*. NY: Gefen, 2007: 64–67, https://www.bio.uci.edu/in-memoriam/howard-maer-lenhoff-ph-d/.

Begin, steeped in Jewish history, knew about Rabbi Reines. Streets in Jerusalem and Tel Aviv are named after the turn-of-the century legendary figure. Lenhoff indicated that Fred Reines was an accomplished physicist. Seeing Fred's name at the top of the list and learning that he was related to the rabbi doubtless impressed the Israeli Prime Minister.

The petition helped prompt Prime Minister Begin to initiate plans to bring suffering Ethiopian Jews to Israel. Lacking permission to fly from Ethiopia, thousands eventually trekked surreptitiously by night across the country to an airstrip in neighboring Sudan. After years of preparation, in 1984 the first wave of 8000 Ethiopian Jews finally flew to Israel in what was called Operation Moses. Subsequently, in Operation Solomon another 14,000 Ethiopians were evacuated from the Sudanese airstrip in secret round-the-clock flights by planes dispatched from Israel. A BBC documentary described the venture as "one of the greatest humanitarian acts of the 20th century."[12]

Whether in the furtherance of science or of humanitarian causes, Fred was an enthusiastic advocate.

---

[12] Francine Wolfisz, Dramatic Rescue of 20,000 Ethiopian Jews Retold in BBC Film, *Jewish News*, December 9, 2015, https://jewishnews.timesofisrael.com/dramatic-rescue-of-20000-ethiopian-jews-retold-in-bbc-film/.

Part V

Expressions

Chapter 12

The "Other" Fred: Music, Performance, Musings

Several leading 20th century physicists were accomplished musicians. Albert Einstein said the most joyful part of his life was playing the violin. Edward Teller, a talented pianist, loved to play pieces by Bach and Beethoven. Richard Feynman was a virtuoso on the bongo drums. But none entertained in public more often and with more pleasure than Fred Reines. During much of his physics career he acted, sang, and whistled before appreciative audiences. The Reines Collection at the UC Irvine library contains scores of playbills, program listings, concert announcements, and newspaper articles about one or another performance. He was variously a light opera soloist and a leading character in theatrical productions.

Fred's early performance experiences were with his high school drama club, and then at college where he acted and sang. The chorus leader at Stevens Institute, impressed by his talent, urged Fred to study with a leading voice coach at the Metropolitan Opera. As Fred later said:

> Since, as a student, I could not afford to pay for lessons, they were eventually provided to me free of charge. Between college and graduate school I even thought briefly about pursuing a professional singing career, but ultimately decided against it.[1]

Meanwhile, Fred was performing in student-written productions as well as in loftier classical works. In 1941, at a joint Stevens

[1] Reines Nobel Lecture.

Institute-Barnard College concert, his deep baritone was on display. His breadth of talent was evident in sonorous solos from two dissimilar compositions: Robert Schumann's operatic *Two Grenadiers* and James Johnson's black spiritual, *Stand Still Jordan*.

During his years in Los Alamos after World War II, he sang and acted in dozens of local productions. His classical repertoire included solos in Handel's *Messiah*, Mendelssohn's *Elijah*, and Bach's *Christmas Oratorio*. A devotee of Gilbert and Sullivan, he played "Dead-Eye" in *HMS Pinafore*, "Private Willis" in *Iolanthe*, and "Colonel Calverley" in *Patience*.

Fred Reines (left) as Colonel Calverley in Gilbert and Sullivan's *Patience*. Los Alamos Light Opera Association, 1954. Reines Collection.

Toward the end of 1955, Fred and his family temporarily moved to the Savannah River facility for his neutrino-detection experiments. His responsibility at the nuclear reactor seemingly was all-consuming. Still he made time for at least one singing engagement. Months after moving and prior to completion of the experiment, Fred's name appeared in a local newspaper. South Carolina's May 21st *Aiken Standard* published an article about an installation dinner for new officers of the local chapter of

Hadassah. Fred was familiar with Hadassah, a national women's organization that supports health and education programs in Israel and the United States. His wife, Sylvia, had served as president of the Los Alamos chapter. The article reported on activities of the Aiken chapter and on remarks by the incoming officers. Then it noted the evening's highlight:

> Dr. Frederick Reines delighted the audience with his singing of the "Volga Boatman" and the "Glory Road." He proved his versatility by his spirited singing of a Jewish folk song which he gave as an encore.[2]

Throughout Fred's adulthood, wherever he was living, he was invited to sing at celebratory occasions. He loved to oblige, whatever the venue, whether religious, civic, or academic. One of his favorite experiences was playing Woody, a leading character in the Los Alamos Light Opera Company production of *Finian's Rainbow*, in 1955. The hit musical comedy opened on Broadway in 1947. The theme draws from an Irish legend that a pot of gold lies buried beneath the end of a rainbow. In the play, after gold was found at the end of a rainbow, Woody arranges that it be shared with local needy villagers. In the end he sings the classic, *Look to the Rainbow*. The lyrics suggest that seeking the end of a rainbow could lead to a treasure. The rainbow is, metaphorically, a dream. The ultimate admonition is to never stop pursuing one's passion. Perhaps Fred realized its symbolic reference to his own pursuit of the neutrino.

Fred moved to Cleveland in 1959 to head the physics department at Case Institute. By then George Szell had been conducting the Cleveland Orchestra for 13 years. When Szell took the position in 1946, the orchestra was regarded as regionally proficient. His conducting drew high praise and the orchestra's reputation soared. It soon became recognized as world-class. "Szell's formidable musicianship and paternal authority," wrote music critic Ted Libbey, "achieved what was probably the highest executant standard of any orchestra in the world."[3]

Among Szell's innovations was the 1952 creation of a professionally trained volunteer chorus. Four years later Szell invited Robert Shaw to

---

[2] "Mrs. Shapiro Speaks to Aiken Hadassah Chapter," *Aiken Standard*, May 21, 1956.

[3] Ted Libbey. *The NPR Listener's Encyclopedia of Classical Music*. NY: Workman Publishing, 2006.

join the orchestra as associate conductor and as conductor of the Cleveland Orchestra Chorus. At age 40, Shaw was already a highly regarded choir director. Although the Robert Shaw Chorale, which he founded in 1948, was widely known, he believed that working with maestro Szell would provide new opportunities. Through the years, the Cleveland Orchestra Chorus performed in cities throughout the world. Each member spent hundreds of hours annually rehearsing and performing.[4]

The departure of some members periodically created opportunities for others to join. Prior to the beginning of a season, Szell and Shaw auditioned new applicants. Some were found unsuitable and others were asked to return for a second audition. Upon arriving at Case in the summer of 1959, Fred Reines submitted his application. Soon after he auditioned, Fred received a letter from Shaw: "Dear Mr. Reines: It is my pleasure to inform you that your audition for the Cleveland Orchestra Chorus has been successful, and you are accepted for membership for the season 1959–1960." Shaw noted further: "By virtue of the fine quality of your audition you are being excused from final auditions." The letter provided details about rehearsal time and location and concluded with the hope that "you will enjoy your work with the chorus."[5]

Fred sang with the Chorus for the next three years. A program from the 1961–1962 season lists his name among the bass vocalists in the performance of Berlioz's *Romeo and Juliet*, Dramatic Symphony, Op. 17. The program also highlighted reviewers' comments about performances during a recent eastern tour.

February 6, *The New York Times*: "The Cleveland Orchestra continues to impress as one of the best in America, which means ... one of the best in the world."

February 12, *The Boston Herald*: "That the Cleveland Orchestra is one of the most distinguished in the world should surprise no one."

February 18, *The Washington Post*: "George Szell brought his superlative orchestra to play for us, and showed us how a large amount of very familiar music can and should sound."

---

[4] "Together, the group's 130 members volunteer more than 30,000 hours annually, through rehearsals and performances." https://www.clevelandorchestra.com/about/choruses-and-ensembles/cleveland-orchestra-chorus/.

[5] Reines Collection.

These orchestra performances apparently did not include the Chorus. Still, as suggested by other observers, the status of the Chorus matched that of the Orchestra. "Shaw was recognized as a builder and master of choirs without peer."

Performances by the Robert Shaw Chorale and the Cleveland Orchestra Chorus in the 1960s remain available on CDs and YouTube. Their sound continues to exude power and majesty.[6] Shaw's demand for perfection was evident not only at rehearsals and performances but also in take-home assignments. The singers received reams of typed advisories concerning vocal technique, philosophy, and the musical selections for the next performance. A portion from a memorandum distributed on February 20, 1960:

> We had a remarkably fine rehearsal last night, but we have one enormous problem to overcome — the seven days between one Monday night and the next. [It] occurs to me that there are ways at hand for us to avoid the beginning where we began instead of where we left off.[7]

Shaw suggested "lots of music marking." He urged the members to mark particular cues with a pen including phrasings, segues, enunciation, rests, and releases.

In addition to marking your music:

a. Concentrate on your intonation.
b. Exercise control in dynamics.
c. Memorize! Try singing with a minimum of two glances per page.
d. Look at the conductor.
e. Feel what you are singing.

Another page seemed intended to instill in the singers a sense of sacred opportunity:

> If Time is always being born it is as certainly always dying. And the great artist, it seems to me will be sensitive to the transience and

[6] Cleveland Orchestra Chorus, https://www.youtube.com/watch?v=R9445pOQw7E https://www.allmusic.com/album/gloria%21-music-of-praise-and-inspiration-mw0000386928.
[7] Reines Collection.

mortality of beauty.... It is this sense of music's "haunting mortality," its but too short loveliness, that, whenever it happens, gives to music urgency and mustness, the lean and the muscular.

Shaw's reflections must have enthralled Fred. He saved dozens of them. He apparently believed that Shaw's drive toward achievement was akin to his own. A letter to Shaw at the end of the 1962 season may or may not have accurately reflected Shaw's feelings about his own work, but it certainly revealed extreme feelings — heaven and hell — that Fred's strivings engendered in himself.

Dear Mr. Shaw,

For the past three seasons I have been a member of the Cleveland Orchestra Chorus. It was fun singing and a privilege to associate with people who take their music making so seriously and do so well. I admire your intense and abiding love and dedication to your art — it must be heaven and hell to be so driven.

Unfortunately my all too finite days and nights press so hard upon me that I must leave the Chorus. Perhaps, the fates and you willing, I'll be able to petition for a return at some future time.

Thank you very much.
Frederick Reines

Fred was honored to have been an invited participant, which was a tribute to his musical talent. But his search for extraterrestrial neutrinos and other academic responsibilities were taking more time.

He had sung with one of the finest choral groups in the world. He never again would take on an extended theatrical or musical commitment. His academic and scientific responsibilities left no time for sustained public performances. Still, he continued to entertain at parties and other one-time events.

In the 1960s, Fred's teenage daughter, Lisa, was strumming her guitar. A familiar melody sounded amid the chords, *My Favorite Things* from the musical, *The Sound of Music*. A tape recorder was running and she called out, "Daddy, please whistle." Preserved on a disk, their sweet–sounding duet is part of the Reines Collection at UC Irvine. Fred's musical whistling was no less hypnotic than his voice.

During his adult years Fred no longer kept a diary, but he wrote essays, sentimental letters (usually not for mailing), and poems. His musings were often introspective and sometimes tapped into personal history. LeRoy Price, a physicist at UC Irvine, joined the Neutrino Group in 1980. He co-authored several scientific articles with Fred, who later asked him to help organize cartons-full of memorabilia, including musings that Fred had written through the years. With Price's help, Fred compiled 100 pages of thoughtful selections that he labelled "Rainbows and Things, A Pot Potpourri." The title was borrowed from the name of the physics course he taught to non-science majors, "Rainbows and Things."

Included in the writings was an essay titled "Who Needs Science?" It was drawn from a 1976 commencement address that Fred had given at the California State University, in Fullerton. The presentation reflected his embrace of scientific inquiry as a core value of human activity. An excerpt:

> [Our] power to exhaust the Earth's resources, to foul the air we breathe and the water we drink has reached alarming proportions. Why not, some suggest, declare a moratorium on science and catch up on our understanding of the mental processes, which take place in the human head? ....
>
> [But] the options, the new solutions, the alterations of limitations, the expansion of possibilities which arise from the pursuit of science do not come on command. They derive from the increased depth of understanding which results from the creative curiosity of educated people....
>
> We have no better way to advance knowledge than to give maximum support to those creative souls among us who want to know "why the grass is green," "why the sun shines," "how electrons behave in semi-conductors," "what are the ultimate constituents of matter," and even what questions are in fact meaningful to ask. Questions such as these — asked when no conceivable consequence was foreseen have led to the extensive knowledge and myriad of things which define our present society. The goal of understanding for the sake of understanding has been the wellspring from which our practical world has sprung....
>
> Who needs science? In sum it is vital to man's material needs, inextricably intertwined with man's search to understand himself and his place in the universe.... Science, a means to practical tools is at the same

> time an activity of transcendent intrinsic merit. It has given us a gift of unlimited value, an endless frontier.

Another vignette, undated, is titled "What Have You Done for Me Lately?" It is about "the combination of insatiable desire and short memory that characterizes us all." Fred recalled that a major fission test under his responsibility was threatened with cancellation because a crucial vacuum system had begun to leak. His reaction revealed a sense of desperation couched in humor. "My concern took the form of a verbal commitment to join the Church if the system held. It did and I haven't."

> Painful though we may find the short-lived nature of gratitude to be I am not convinced it is such a shortcoming. If we rested on our laurels we would wither and decay. If we did not continually have to renew the other fellow's approval of our actions on his behalf, the glue that binds us together would lose its strength. Of course this sort of thing can be overdone. We can spend all our time trying to make the other fellow beholden to us or in trying to satisfy ourselves by reaching for goals beyond any hope of attainment.
>
> I occasionally yearn for some measure of contentment in this unbalanced struggle for approbation both by my fellow man and myself. I think we all do and, you know, I doubt if we'll ever find it — and I confess I really hope we don't.

As Fred reread the collected material, Price observed him laugh while reviewing one essay in particular. It was about an experience when Fred was a Boy Scout. Like Fred, Price had attained the rank of Eagle, an accomplishment they both savored. Titled "Lizzy Diving," an excerpt expresses its essence:

> About a half century ago I was a Boy Scout — an Eagle Scout no less, imbued with the lofty maxim of the Scout code — a scout is trustworthy, loyal, helpful, friendly, courteous, kind, obedient, cheerful, thrifty, brave, clean and reverent, and one who loves his country and its heritage.
>
> Well-armed with all this noble tradition and believing it all as simple, clear, and true, I decided in the full wisdom of a 16-year old that I wanted to do a man's job that was important to society, a job that would

exemplify honest and honorable toil, a job that would use my budding mechanical skills — a job that would be appreciated by my fellows. I had the good fortune to find one in the camp where I had completed the requirements for the Eagle Scout Badge!

It was the job of Lizzy Diving, that euphonious title which labelled the keeper of the latrines.

Here, reasoned the newly minted Eagle, was a real challenge. He was to be the Sanitary Engineer for the entire camp of 300 healthy boys, his task to maintain the gasoline pump supplying the camp with water and with keeping the latrines clean and in good operating condition. The odds were enormous — 300 to one — matching the young appetites and incredible through-put.

During my month of stewardship I would enter the wooden house, a four holer, and make a professional survey of the disaster area created by my fellow scouts. I could tell you how it was necessary to be extremely agile if one was to avoid the inevitable splash as the plug was removed beneath the quartet of thrones and the contents started to flow to their final resting place, the septic tank. I could tell you how I hosed out the facility and washed all with sweet smelling pine oil. I could tell you all these things and many more, but wish to spare you such unpleasant tales.

But I *will* tell you that though I wouldn't prefer it to a picnic, it wasn't nearly as onerous a task as one might imagine. It was an experience for which I am grateful — it taught me the dignity of honest toil and the importance of meeting worthwhile ends.

During an interview in 2018, emeritus researcher LeRoy Price recounted Fred's pleasure in reading the essay. He also revealed his own pleasure in having worked with Fred:

> Not only was Fred a brilliant and imaginative physicist. But there were so many times at meetings, maybe a dozen people in his group would be discussing something and Fred would say, "Well what about 'such and such', something very fundamental, and everybody in the room looked at each other as if to say, "Why didn't I think of that? It is so obvious, so important." He had this knack of picking out the truly important points from a mix of ideas. That was a talent that I absolutely admired.[8]

[8] LeRoy Price interview, 2018.

The lobby of Reines Hall, the Physical Science building at UC Irvine, is filled with memorabilia about Fred. Showcases include dozens of poems that Fred composed through the years, many of which appear in the Reines Pot Pourri pages. They range from the whimsical to the somber. A sample:

Viva la Difference

Curves figure in sex
In manner complex
The impulse, the action
The core of attraction
Is concavity
Meeting convex.

To a Revolutionary

Tell me, my starry eyed idealist
What will you do when the revolution
  comes and is past?
Will you be content in the world you
  made or will your keen discerning eyes
Discover more evils to uproot with force.
Cast in the mold of fomenter of change,
Can you live long in a comparatively
  static world?
No, there will always be another
  revolution for you.
You will find no peace for you will
  be a pariah in the land of your own making.
How strong your motives and how terrible your means!
How like your means are your returns!
Is martyrdom part of your plans?
If not dear revolutionary,
  come sit on the fence with me.

A Toast

Here's to Wolfgang Pauli
  who made a funny joke
Here's to the Great Enrico

who then of weakness spoke
Here's to all those gathered
to celebrate the fruits
Of all the patient workers
who followed those astutes
Here's to the proposition
That we shall meet again
And Here's to the fond hope
The sun will shine till then.

MAD [Mutual Assured Destruction]

What brought us to the brink
To the very edge of the abyss?
To the place where the fires of Hell
Are seen with threatening clarity?

How came we to this state
Where in the twinkling of an eye
Humanity, its nobility awry,
Might vanish?

What brought us to the brink?
How, step by logical step
Have we marched toward the abyss
To the very edge of doom
To the end of man's adventure
How came we to this state?
Can we, searching, find our way out of this damned maze
toward an earth suffused with love
to a brotherhood of man —
Somehow divine a cosmic plan
Where individuals lend to one another a
helping hand and collective idiocy is banned
Where peace prevails and Hell's fires
no longer blind out eyes?

Fred's singing and acting, his essays and poetry, his scientific accomplishments, reveal his far-reaching interests and talent. His mind remained restless, his curiosity unflagging.

Chapter 13

Science and Religion

A Google search for "science and religion" lists dozens of sites. It also mentions subcategories including "science and religion coexist," "science and religion conflict," and "similarities between science and religion." As suggested by the titles, the two fields are variously viewed as compatible, irreconcilable, and similar. Fred Reines thought, wrote, and spoke about the relationship. For him science usually came out as an unequivocal good, while about religion he remained ambivalent. He began pondering the essence of each in his teen years and continued to do so through adulthood.

At 16, Fred decided to maintain a daily journal. He had already been wrestling with ideas about the world and his place in it. His initial entry, in 1934, revealed his determination to summarize and organize his thoughts. Recording them regularly on paper, he believed, would demonstrate self-discipline and enhance self-understanding.

> Today, Dec. 8, I have made my mind up to keep a diary. This resolve was precipitated by [a] new notebook and the pleasure I will receive as a reward if I have the willpower to continue with this task. Herein every night before retiring, tired or full of energy, I shall make entries of not less than a page, of the happenings of the day, be they prosaic or novel. I feel that I need something to sound out my thoughts and attitudes on various subjects. Also have I felt the need for some backbone of my activities. These shall my diary fulfill.[1]

[1] Reines Collection.

An early entry — on connections and contrasts between religion and science — addresses a topic he returned to frequently. His diary revealed an adolescent with questions about his own religious and ethnic identity.

On January 16, 1935 he wrote about the impact of *The Island Within*, an autobiographical novel by Ludwig Lewisohn. Published in 1928, the book had reached *The New York Times* Best Seller list. Its main character is Arthur Levy. Like Lewisohn, Arthur was the son of assimilated German Jews who came to America when he was a child. As Arthur grew up, he tried to shed his Jewish identity, married a Christian woman, and sought friendships mainly with non-Jews.

Mirroring Lewisohn's life, the fictional Arthur ultimately felt discomfort with his own assimilation efforts. Lewisohn had been denied job opportunities because of his Jewish lineage, and he had experienced anti-Semitic barbs from "friends." His wife's attachment to her Protestant heritage prompted him to engage with his own roots. Lewisohn's marital dissonance (like Arthur's) led to divorce. His real-life trajectory paralleled that of his fictional doppelgänger.

Until Arthur's divorce, according to a reviewer of the book, his life was "dominated by his futile struggle to free himself from the constraints of his ethnic background."[2]

Subsequently Arthur (like Lewisohn) finds salvation in his Jewish identity: in recognizing the shared history and the struggles of the Jewish people. Their attachment was driven less by religious orthodoxy than a sense of peoplehood. Finding security would require Jews to acknowledge their heritage and to unite. As Lewisohn came to believe, this could only be accomplished through Zionism, based on the millennia-long cultural and religious ties of the Jewish people to the land of Israel. Destined to be outsiders in other communities, he championed the re-establishment of a Jewish state in their ancestral homeland.

The book had a profound effect on Fred. He evidently put other things aside to read it. His diary entry began:

> Today, in school, and at home I read Lewisohn's *Island Within* to completion. After which at 1:45 a.m. I retired.

---

[2]Hedge N. Nilsen, "Jewish Nationalism: A Reading of Ludwig Lewisohn's *The Island Within*" https://rauli.cbs.dk/index.php/assc/article/download/1202/1201.

Fred was transfixed by Arthur Levy's shift from rejection of his Jewish heritage to embracing it. Young Reines was growing to maturity at a time when anti-Semitism was rife (which is still the case in much of the world). Treatment of Jews was becoming especially barbaric in Germany under Hitler. But prejudice against Jews in America, if less savage, was ubiquitous. Institutional anti-Semitism was a cultural norm. Jewish communities built their own hospitals because Jewish physicians were often denied affiliations elsewhere. Similarly, they had to establish their own law firms and financial institutions. They were excluded from management positions in major corporations and in government agencies. Student applicants to leading colleges were subjected to "Jewish quotas." Faculty positions were often denied to Jews.

Some young Jews reacted defensively as they became aware of this reality. They pondered whether disavowing their Jewish identity would save them from discrimination. Some even internalized disparagement of stereotypically Jewish behaviors. That was true for Arthur Levy, at least initially. But Arthur's ventures also showed a way forward, which affected Fred:

> Lewisohn's book, full of familiar references to Jewish characteristics, has again made me feel that I am a Jew. There seems to be no getting away from that elementary fact. In fact I am not certain that I would be happy if I did succeed [in rejecting that fact].

Fred wondered whether his own "dislike [of] some characteristics" ascribed to Jews was not the result of conditioning by the larger society. His observations prompted me to read a copy of *The Island Within*. I found a myriad of allegedly Jewish characteristics that Fred must have been alluding to. The contradictory mix was echoed by the story's different protagonists. Jews were variously labeled: Learned. Wise. Cunning. Vulgar. Creative. Astute. Sensitive.

Lewisohn reflected on the quandary posed by such a list: "No wonder people changed their names and tried to escape so cruel and absurd a dilemma." Through Arthur's voice he proposed that "Science could be salvation.... To know the causes of things, to know things undeludedly and in their true character. Thus one transcended the follies of prejudices of the herd."[3] Fred's attitude about science was already in sync with these sentiments as his diary makes clear:

> I shall attempt to become a scientist so that I may live in a world where racial discrimination, [and the] ignorant and unjust, can never enter. This resolve, which has influenced my life since I began [this] understanding a few years ago, was strengthened by the actions of Arthur Levy, who strayed far but then came close to the faith.

Striving for scientific "truths" is a worthy endeavor. But believing that science and scientists are the key to a world devoid of prejudice seems naïve. Young Fred's struggle to resolve conflicting sentiments is palpable in a diary entry that alternates praise and disdain for an obviously bigoted "friend."

> This afternoon, I walked home with Hans who spoke to me about a book he is reading. I received the impression that he is an excellent reader and raconteur. Strange. He is a bundle of prejudices, with many regarding Jews, and yet he isn't a bad fellow. He talks fairly intelligently some of the time — but he is an exhibitionist (with only dumb statements and a crazy laugh to exhibit).

Despite fulfillment of his ambition to be a scientist, Fred never ceased weighing the place of religion and divinity in human affairs.

Decades later Fred wrote an undated essay (probably in the 1970s) titled, "Some Thoughts on Religion."[4] He recalled a night-long conversation about religion that he had with Clyde Cowan, his colleague at the Los Alamos National Laboratory. At the Nevada test site in 1951 or 1952, they were preparing for a nuclear test the following morning. Cowan was a devout Catholic. Fred began with a provocative premise.

> Since I liked him so much I engaged in a rationalization of his belief that Jesus is the Son of God. I said if the characteristics ascribed to Jesus as

---

[3] Ludwig Lewisohn, *The Island Within*, NY: Harper and Brothers, 1928. Reprinted by Syracuse University Press, 1997: 117, 166.

[4] Reines Collection.

> a man — namely humility, love of his fellow man, purity of heart — won him this appellation then [others who] possessed these traits, also qualified for the title.

Fred's position was that humans are a subset of all creation and that all persons are therefore "sons (and daughters) of God." He did not record Cowan's response, though it presumably reflected the precepts of Clyde's Catholic faith. Cowan died in 1974 and subsequently his widow, Betty, and their son George spoke about his attachment to Catholicism. They did so on a program sponsored by the Knights of Columbus, mentioned in Chapter 8. Clyde's comments to Fred may be surmised by Betty and George's observations. They both alluded to his deep faith and the fact that his beliefs were framed by Catholic precepts. This would include belief that Jesus was the son of God and that the Bible is the error-free and revealed word of God.

How did this square with science? George recalled his father saying that science enhanced his faith. He remained enchanted by the notion that solving a scientific mystery would commonly result in facing more and perhaps greater mysteries. Fred later wrote about his lengthy discussion with Cowan at the Nevada site:

> Clyde and I emerged from the night long vigil with our pre-discussion points of view intact. As I recall we made a tacit agreement to avoid such irreconcilable topics in the future and stick to our physics. Having accepted this constraint our friendship flourished and we were privileged to climb a few mountains together.

Fred's essay, "Some Thoughts on Religion," also revealed an ongoing effort to understand what he called "the scheme of things." He expounded on such concepts as the "workings of nature" and "divinity."

> Some people are privileged to see more deeply into the workings of nature than others; if they do physics they are called great physicists; if they do medicine they are called great physicians; if they try to raise the condition of man they may be called Prophets.... As men differ one from the other so some have the greater vision — so some say we are divinely inspired. The great actress Sarah Bernhardt was

such a consummate artist she was accorded the appellation "divine." And the list of greater and lesser "divinities" can be as long as you care to make it.

In 1959, his last year at Los Alamos, Fred's stage performances ended with a role that he relished. The local theater group performed *Inherit the Wind*, a play that opened on Broadway four years earlier. It was based on the famous 1925 "Monkey Trial" in Dayton, Tennessee. Fred played Henry Drummond, who was a depiction of Clarence Darrow, the great civil rights attorney. No role could have better reflected Fred's own views: Belief in the absolute rationality of science, but ambivalence about the proper place of religion in human affairs.

Clarence Darrow is interpreted by the character Drummond, played by Fred Reines. Reines is world famous for his codiscovery of the neutrino. His stage experience includes the romantic lead in "Finian's Rainbow," and many appearances in the Gilbert and Sullivan productions by the Light Opera Association of Los Alamos. Most recently he appeared in the local production of "Arms and the Man."

Local newspaper account of Fred's performance in *Inherit the Wind*, 1959, Reines Collection.

The core issue of the actual trial was whether human origins were attributable to evolution or to literal biblical descriptions. John Scopes, a local teacher, had been charged with violating a state law that prohibited the teaching of evolution in public schools. Scopes admitted teaching the subject, but denied culpability, claiming that evolution was scientifically valid. The trial's significance was underscored by the participation of Darrow and William Jennings Bryan, a renowned former statesman and three-time (unsuccessful) presidential candidate. Bryan would lead the prosecution on behalf of the state. He characterized evolution derisively as holding that humans descended from monkeys. The Bible was the word of God and Scopes had broken the law.

Darrow argued for the defense. He maintained that the law was based on unscientific religious beliefs. The charges against Scopes, he contended, were an unconstitutional abridgement of the teacher's freedom of speech.

The trial judge would not allow the defense team's scientific experts to testify. He ruled that their purported expertise was not germane to the case. A frustrated Darrow then proposed that Bryan himself be called as an expert witness on the Bible. The judge acceded, and Bryan submitted to questioning by Darrow.

In the end Darrow was generally seen as having bettered Bryan and his fundamentalist convictions. Still, the jury convicted Scopes and the judge fined him $100. The conviction was later overturned on a technicality. The judge had instructed the jury to consider only whether Scopes had broken the law, not whether the law was just.

Because the play was partly fictional, Jerome Lawrence and Robert Lee's *Inherit the Wind* changed the names of the trial's protagonists: Bryan became Matthew Harrison Brady and Darrow became Henry Drummond. They acknowledged writing some of the script for theatrical effect.[5] When the Los Alamos theater group performed *Inherit the Wind*, Fred must have been thrilled to present Drummond's scientific arguments. The play's denouement occurred when Brady (William Jennings Bryan) was in the witness chair:

> D (Fred): Now tell me. Do you feel that every word written in [the Bible] should be taken literally?
>
> B: Every word in the Bible should be accepted exactly as it is given there.
>
> D: Now take this place where the whale swallows Jonah. Do you feel that actually happened? Matter of fact it says "a great fish" — but it's pretty much the same thing. What's your feeling about that?
>
> B: I believe in a God who can make a whale and who can make a man and make both do as He pleases....

---

[5] Jerome Lawrence and Robert E. Lee, *Inherit the Wind*, NY: Random House, 1955.

D: Have you ever pondered what would naturally happen to the earth if the sun stood still? ....

B: I have faith in the Bible.

D: You don't have much faith in the solar system.

B: The sun stopped.... Natural law was born in the mind of the Heavenly Father. He can change it, cancel it, use it as He pleases. It constantly amazes me that you apostles of science, for all your supposed wisdom, fail to grasp this simple fact.

D: Listen to this: Genesis 4:16. "And Cain went out from the presence of the Lord, and dwelt in the land of Nod, on the East of Eden. And Cain *knew his wife*!" Where the hell did *she* come from?

B: Who?

D: Mrs. Cain. Cain's wife. If, "In the beginning" there were only Adam and Eve, and Cain and Abel, where did this extra woman spring from? Ever figure that out?

B: No, sir. I will leave the agnostics to hunt for her.

Drummond cited additional biblical passages that were irreconcilable with scientific reality or even with common sense. Brady continued to insist that the Bible was holy and its text unchallengeable. The exchanges became acrimonious. The courtroom spectators had become unruly and the judge adjourned the session.

The final act included the jury's verdict against the play's Scopes character. Drummond promised to appeal and assured him that in any case he had already won. "Millions of people will say ... that you smashed a bad law."

In a final twist, despite Drummond's harsh examination of Brady during the trial, he later paid respect to Brady's contributions to America. "There was much greatness in this man," Drummond said. "A giant once lived in that body. But Matt Brady got lost. Because he was looking for God too high up and too far away."

That gesture was consistent with Fred's own values. In the world of science, competition among researchers could be fierce. Disputed primacy about discoveries has at times been contentious. Fred had never allowed professional disagreement with others to lessen respect for their achievements.

In subsequent years, Fred would continue to speak about science and religion. In 1974, he was invited to talk on the subject to the Women Associates of the University Interfaith Foundation at Irvine. In notes for the talk he wrote:

> It is the dogma of science to assume the universe is knowable. It is the wonder of science that we have come as far as we have. So you see at the base of science is faith, a faith in my view somewhat akin to the faith of theology.

Fred's notes also refer to a debate he once had with a Dominican Father. The priest was not happy.

> He was most disapproving of my thesis that the universe was knowable by rational means. He said I had too much faith!

His talk included a topical heading in the form of an underlined question: What is Science? His written answer:

> It is the search for the eternal verities. In a complex world of diversity and change what are those things that remain constant? The success of science in advancing our understanding of the physical universe lies in the discovery of conservation laws.

Fred saw human history as drenched in superstitious beliefs. The space for these beliefs is continuously reduced as science advances.

> Prior to the development of modern science we deified our ignorance. If it thundered we created a God of thunderbolts. We filled the seas, the skies, the forests, the mountains with spirits, wizards, demons.... Science proceeded to exorcise the demons — the lightning became electric currents produced by the separation of charge; the thunder, the air rushing into the partial vacuum created by the electrical discharge. The universe became a more rational place. Mystery was replaced by a measure of understanding.

He also drafted a question in his preparatory notes that presumably he raised before the Interfaith Foundation audience: Would all physical mysteries eventually be solved by scientific understanding? He believed this was possible, but it still does not explain why nature works as it does. The answer to this question, he said, "is the ultimate goal of science."

Fred's belief that science could eventually reveal the underpinnings of all physical mysteries is truly a matter of faith. His idealization of science was rooted in his teenage supposition nearly 40 years earlier that science could also end discrimination and injustice.

Chapter 14

Classroom Teacher

In June 1972, Fred Reines received a letter from Dr. M.J. Renan, a physicist at the University of Witwatersrand in South Africa:

> Dear Professor Reines:
>
> I believe that you are responsible for a very successful and well received Physics course, aimed primarily at the non-science major student, at the University of California, Irvine. I should be very grateful if you could send me a few details concerning this course, since I shall be handling a similar series of lectures at our University. Information such as prescribed textbooks, plan or outline of course, or perhaps even some duplicated lecture material, would be particularly welcome.
>
> With thanks,
> Yours sincerely,
>
> M.J. Renan
> Research Officer

Fred's comprehensive response began with a statement of aims for the course:

> Dear Dr. Renan:
>
> Thank you for your inquiry regarding my physics course for non-science majors (I call it "Rainbows and Things")....
>
> My main objective is to introduce non-scientists to an awareness in a participatory manner of the fact that the scientist's view of the universe

> is derived from the commonly held (at least by the very young) curiosity regarding the world around him/her. I try to do this by showing them that science is not a thing apart from poetry, philosophy, the arts, but stems from the same roots. It represents but one facet of man's creativity, but one which has the possibility of leading in a distinctive manner to ask questions of nature and, surprisingly often, to receive answers. I try to tell my students that it is sometimes possible in science to maintain a degree of objectivity that is often denied other fields of human endeavor. I try to tell them that this habit of objectively weighing the facts is, joyfully, contagious and might even help them to construct a more rational world.

Fred's manner of teaching was innovative and flexible. His letter included a list of techniques and observations, which he termed his modus operandi. Following is an edited version of the list:

> (1) The class is on a pass/no pass basis. There are no other grades given.
> (2) Two essays completely of the student's own choosing are required, one in mid-course, the other at the end.
> (3) There are no prescribed texts although I refer them to very many. I urge them to discover the library.
> (4) Such a class is best taught to a smallish number, say 25, [which] was the largest number I could deal with as individuals. Unfortunately I understand that next year I must teach a few hundred, and a dialogue will then become a performance.
> (5) Experience shows that in such a 10-week course (two separate hours each week plus a multitude of discussions with the students) it is advisable to devote the first half of each hour to carrying forward the thread of the subject and the remainder to answering questions. This can be tough on the instructor but the students like it.
> (6) In discussing physics one should eschew any mathematics beyond the most elementary algebra and give experimental demonstrations using only the simplest equipment, (e.g. I use a candle, a piece of cardboard with a pinhole in it, a large lens, a piece of string, a paper clip. I whistle along with a student to illustrate beat frequencies, etc.). The object is to sharpen the student's interest in doing some experiment him/her self. The results are sometimes absolutely incredible.
> (7) It is possible to start with the question of "What is a Rainbow?" read poetry, and lead gradually to wave particle dualism and the

essential ideas of special relativity. (Incidentally, I regard it an absurdity that there are so many purportedly educated people who haven't the slightest idea of these exciting and revolutionary aspects of our intellectual and practical world!)

(8) Some of the essays evoked are truly remarkable. I have received the gamut from poetry, playlets, watercolors, to detailed scholarly experimental studies on soap bubbles, holography, tachyons, demonstrations of the wave nature of light, autobiographical sketches, photographs taken underwater to demonstrate total internal reflection and how the world appears to a fish (!), photographs showing the changing shape of the sun as it sets.

(9) The students are predominantly freshmen with a sprinkling of sophomores and the upper classes.

(10) Unfortunately I have no notes in a form sufficiently intelligible to anyone but myself so I must ask you to excuse me for not sending them to you.

I hope these remarks are of some help to you and I would like to add a word of admiration for your interest in teaching science to non-science majors. We in science have an obligation to share our sense of wonder with the rest of our fellows. More practically, why should they continue to support our studies of science when they are not given a clue that it is comprehensible and touches their lives in many ways ranging from the quality of intellectual life to the electricity that lights our homes in the darkness.

Please when you see them give my regards to [Witwatersrand] Professors Sellschop and Nabarro.

Sincerely yours,
F. Reines
Professor of Physics

Fred's letter reflected both his devotion to science and a sense of responsibility to promote it to others. As mentioned previously, his fascination began late one afternoon when he was a youngster. While squinting through his curled fingers toward a window, he saw a peculiar light image that he later understood to be the phenomenon of diffraction.[1]

---

[1] Nobel lecture.

Diffraction can reveal white light as a composite of colors, much as refraction can create a rainbow as sunlight passes through water droplets. The colors are determined by their different wavelengths. They become individually visible when light is bent by traveling through a narrow slit — diffraction. Or when sunlight passes through droplets that act as prisms — refraction.

The university's academic year was divided into four 10-week quarters, and beginning in 1970 Fred regularly taught his quarter-long course. About two dozen students signed up when he first offered the subject, but the popularity of his class grew quickly, and by 1973 it was drawing more than 200 enrollees. In Fred's letter to Renan he characterized his teaching approach as having moved from dialogue when classes were small, to performance, when they grew larger. His years of stage acting and singing assured a compelling presence no matter how large the class.

The Reines Collection at UC Irvine includes discs made from audio tapes of his initial classes. Fred's manner of speech was assertive. He welcomed questions from students, and he answered crisply and respectfully. Listening to the recordings, I heard a voice that had changed little over the years. It remained deeply resonant and his delivery, never rushed, was offered with the cadence of a skilled performer.

My inability to view his demonstrations and blackboard notations limited a full understanding of his presentations. Still, it was obvious the course design was innovative and flexible.

Fred began his first class by stating his expectations of the students. He hoped they would be inspired by readings and class discussion to perform some simple experiments on their own. By the course's end they would be required to submit some essays. The essays could be on scientific experiments or on poetry or philosophy based on what a student had read or what was discussed in class. He then recited several passages from *The People, Yes*, a book-length poem by Carl Sandburg, which Fred described as "pretty remarkable." An excerpt:

> A father sees a son nearing manhood.
> What should he tell him?....
> Tell him to be a fool every so often
> and to have no shame over having been a fool
> yet learning something out of every folly

hoping to repeat none of the cheap follies
thus arriving at intimate understanding
of a world numbering many fools.
Tell him to be alone often and get at himself
and above all tell himself no lies about himself
whatever the white lies and protective fronts
he may use amongst other people.
Tell him solitude is creative if he is strong
and the final decisions are made in silent rooms.
Tell him to be different from other people
if it comes natural and easy being different.
Let him have lazy days seeking his deeper motives.
Let him seek deep for where he is a born natural.

Then he may understand Shakespeare
and the Wright brothers, Pasteur, Pavlov,
Michael Faraday and free imaginations
bringing changes into a world resenting change.

He will be lonely enough
to have time for the work
he knows as his own.[2]

This passage appeared to mirror Fred's own behavior and stated values — take risks; learn from your mistakes; be creative; be different; bring changes into the world. Sandburg's inclusion of great names in literature, invention, and the sciences, also reflected Fred's high regard for creativity, whatever the discipline.

After reading the poem to the class he pronounced what he called the most important part of his introductory remarks. He urged the students to question freely and not worry about appearing uninformed. He will make suggestions, and their reactions and questions will help guide the direction of the course.

He veered to a discussion about rainbows. After a recent heavy rain he had seen an enchanting series of rainbows that reminded him of a line by the 19th century English poet William Wordsworth: "My heart leaps up

[2] Transcript in Reines Collection, http://www.yourdailypoem.com/listpoem.jsp?poem_id=1501.

when I behold a rainbow in the sky." The phenomenon, Fred said, is the stuff of legends:

> There is the story of the pot of gold at the end of the rainbow and the leprechaun who has the key to the pot of gold. There's a whole legend about this — the hope, the inspiration. You see a rainbow in the sky and you say: Things are looking better now.

Fred's familiarity with this legend, although not mentioned in the classroom, had been sharpened by his stage performance a decade earlier in *Finian's Rainbow*.

He went on to note that rainbows are visual images that "we clothe in poetic, philosophic legendry." Behold the beautiful arcs of red, orange, yellow, green, blue, indigo, and violet. But to truly understand a rainbow requires understanding its relationship to other things: to water, to light, to wavelengths. "Understanding relationships is what science is all about," he said.

Fred had seen people pause to admire a setting sun turned red. He suggested that the class should understand the physical basis of the appearance. He rejected the notion that understanding a sunset diminishes a sense of its beauty. He declared that, "you don't like something less or have less emotion about it by understanding it. You don't make the thing uninteresting because certain aspects are a mystery, because you dispel some ignorance about it."

At sunset and sunrise the sun's light travels through more atmosphere than during mid-day. The increased number of atmospheric particles deflects light of shorter wavelengths. Red wavelengths are the longest of all the colors' wavelengths that together compose white light. The red wavelengths are the least scattered and thus remain visible when the sun is near the horizon.

As Fred could avow from his own experience, knowing that the sun's red color was a consequence of waves, refraction, deflection, and the position of the viewer in no way altered its beauty.

Fred's approach to teaching may be gleaned from an audio recording of an interaction with some students. The subject was the refraction of light through a prism and the resulting rainbow effect.

> Reines: At a certain angular position the light will reflect what color I will have. [Fred evidently was demonstrating a position as he spoke.]

Here's the eye — What color will I have? Red, that's what I'll have. If I pick a different angular position I'll have different colors. How about that?

Student-1: That means depending on the position, you get different colors?

R: Depending on where I look. If I didn't look up I wouldn't see a damn thing.

S-1: If it's just one drop I can see that. But when you have a lot of water droplets it seems like they would be refracting one light down there and another over here. Why don't they combine and make a white light coming to you?

R: Because the different colors come from different directions. You're quite right that if all the colors came at the same angle you'd have white light again, and no rainbow. Then we could put our poetry away.

Student-2: When we see a rainbow and it's not raining where you are, it's that the rain is falling where the rainbow is?

R: That's right. The rainbow is something that you can never reach. The rainbow is a very personal thing because each person sees his own rainbow.

S-2: You mean that if two people are standing together then each one of them will not see the same rainbow?

R: Well, if they're close enough together and have the same angles they'll see it. But if some joker is over there, he'll say, What rainbow?

S-2: So everyone sees a different rainbow?

R: Let's suppose we get into this more deeply. The moat, as they say, is in the eyes of the beholder. What people see or hear depends in large part on what they are prepared to see or hear. If it is a Bach or Vivaldi transcription and one's ear is not attuned to that kind of music it will sound like garbage. But if a person is attuned to it, he will be inspired.

A week later another student said that after the last class he tried an experiment. "I took a thick piece of plexiglass which is transparent and shone a light at different angles through it, but I didn't see any coherent patterns." Fred responded:

Well the trouble is that plexiglass isn't perfectly uniform. See, when I take a bubble, that is the most uniform thickness you ever saw in the world, so it's just the precise number of wave lengths and what have you.

> But this damn plexiglass has got wavelengths on it that are many, many wavelengths different. So there's no coherence. It's got to be perfectly regular or it's no dice.

In subsequent class sessions Fred discussed a range of physics subjects including, as he wrote to Renan, wave-particle duality and special relativity. The duality refers to the phenomenon of light and matter exhibiting both wave and particle properties. Special relativity was conceptualized in 1905 by Albert Einstein. He posited that the speed of light is constant no matter the position or motion of an observer.

In 2004, *The Chronicle of Higher Education* published an article titled, "What Makes Great Teachers Great?"[3] The author, Ken Bain, was among 60 professors from various disciplines who examined the qualities that outstanding teachers have in common. The most important characteristic, they concluded, was the creation of "a natural critical learning environment."

> In this environment students learn to reason from evidence, to examine the quality of their reasoning, and to ask probing and insightful questions. They are encouraged to compare, apply, evaluate, analyze, and synthesize, but never only to listen and remember.

Bain contrasted this with less desirable methods of instruction:

> Many teachers never raise questions; they simply give students answers. If they do tackle intellectual problems, they often focus only on their subject and the issues that animate the most sophisticated scholarship in the field. In contrast, the best teachers tend to embed the discipline's issues in broader concerns, often taking an interdisciplinary approach.

Fred's course objectives and teaching technique in Rainbows and Things mirror the ideals expressed in Bain's article. His approach was appreciated in higher level physics classes as well.

---

[3] Ken Bain, "What Makes Great Teachers Great?" *The Chronicle of Higher Education*, April 9, 2004.

Teaching physics, Reines Collection.

In 1990, Christine Chen was in her third year of high school. Her father Herbert Chen had been an esteemed physicist at UC Irvine until his premature death three years earlier. Christine was disappointed that her high school offered no physics courses. An excellent student, she was able to take basic physics at the university, whose campus was a 10-minute walk from her high school. The curriculum consisted of 10-week quarters, each quarter covering a different aspect of physics, that extended into her senior year of high school.

The physics subjects included mechanics, electricity, magnetism, waves, thermodynamics, and more. Altogether they comprised the core curriculum for undergraduate physics majors. Professor Fred Reines taught the section on waves. Christine recalled that when she took his course, "lots of people knew that he loved to sing opera and that he was quite the performer." She found him "very entertaining."[4]

A sound wave may be depicted as a sequence of peaks and valleys. Over time, the wave line curves up and down at a certain frequency. "Beating" is a physics term that describes the effects of two sound waves with slightly different frequencies playing at the same time. When the peaks match each other the sound is louder, and when they don't, it is softer. The resulting tone is wobbly: loud-soft-loud-soft. The phenomenon is commonly demonstrated by sound-producing devices, like a tuning fork or musical instrument. But not in Professor Reines' classroom, Christine said.

> I remember once when he was trying to explain beating, when you have two frequencies that are slightly different. There is a periodic signature between them where they get "brighter" depending how they synchronize with each other. Fred invited a student down to the front of the class and they both needed to whistle, though at different frequencies, to illustrate the beating frequency. He was a fun lecturer and used real world examples like that to illustrate concepts.

On July 11, 1991, southern California experienced a partial solar eclipse. The sky darkened for 6 minutes that morning as the moon passed between the earth and the sun. Since the eclipse was not total, a sliver of light remained visible at the sun's periphery.

When the eclipse occurred, Christine was in a lecture hall filled with students whose eyes were fixed on a large white wall. "Somebody had put up this telescope where they could project the solar eclipse onto the wall so that everybody could see it," Christine said. Suddenly laughter broke out from the roomful of viewers. The silhouette of a rabbit had appeared on the sun's lit rim and the rabbit's ears began to wiggle as its head bobbed in different directions.

The image was a shadow puppet. Fred Reines was manipulating his hands and fingers in front of the projected eclipse. It was another playful display that endeared him to Christine and his other students. Christine went on to become an astronomer. She currently is with the Space

[4] Christine Chen interview, 2018.

Telescope Science Institute in Baltimore and is a research scientist in the Department of Physics and Astronomy at Johns Hopkins University.

Christine Chen, personal website, https://www.stsci.edu/~cchen/STScI/index.html.

Fred enjoyed teaching and mentoring while continuing neutrino research. But he also felt that recognition of his professional achievements had lagged. His three siblings — Bill, Dave, and Paula — were immensely proud of him, as had been their mother until her death in 1972. They were also aware of Fred's expectation that his detection of the neutrino would be rewarded with much acclaim.

For some period after the detection, as Bill and Dave told me, Fred was disappointed that his discovery had brought him less recognition than he had anticipated. Not only was he being passed over by the Nobel committee, but even lesser awards had been infrequent. In the 1950s and 1960s he was named a "fellow" by a few groups, including the American Physical Society and the American Academy of Arts and Sciences. He received an honorary degree from the University of Witwatersrand and in 1969 was inducted, belatedly, into Phi Beta Kappa at Stevens Institute. Then, nothing during the following decade.

The trend began to change in 1979 with a fellowship in the American Association for the Advancement of Science. Honors continued to follow including election to the National Academy of Sciences and receipt of the National Medal of Science. Still, no Nobel Prize.

Meanwhile, Reines remained dedicated to teaching, mentoring, and the notion that science was indispensable to a rational life. His efforts to advance science were also expressed by his own work in particle physics, especially regarding neutrinos. His full engagement continued into the 1980s.

Part VI

Validation

Chapter 15

IMB and a Decade of Joy and Frustration

In 1967, during the height of the Cold War, Andrei Sakharov, father of the Soviet hydrogen bomb, wrote two memorable letters. The eminent physicist was increasingly concerned about the risks of a nuclear arms race. The first letter was a secret plea to Soviet leaders to accept an American proposal for a moratorium on developing strategic anti-ballistic missiles. His message was ignored. Sakharov later openly criticized Soviet policies on nuclear security and human rights. As a result, he was banned from classified research and from making any public display of his dissenting views. His repression ended in the late 1980s after Mikhail Gorbachev became the Soviet premier. Sakharov's vindication was complete when the Soviet regime collapsed in 1991.

Unlike his secret memorandum to the Kremlin rulers, Sakharov's second message appeared in the prestigious Soviet *Journal of Experimental and Theoretical Physics*. While the issues raised in the first letter ended with the collapse of the Soviet Union, those in the second remain unresolved. Their effect continues to reverberate among scientists who seek to understand how the universe came to exist.

In 1911, Ernest Rutherford discovered that a tiny positively charged ball — a nucleus — sits at the center of every atom. He later understood that the nuclear charge is attributable to a particle that he called a proton. Protons inhabit the nuclei of atoms, and in electrically neutral atoms their positive charge is balanced by negatively charged electrons outside the nucleus. Even when not bound in an atom, a free proton — of which there are many in stars and in "empty" space — was believed to be stable: that is, to have an infinite half-life (i.e. they live forever).

James Chadwick's discovery of the neutron in 1932 led to a new understanding of the essential nature of an atom. Like protons, neutrons reside in atomic nuclei, though they don't contribute to the nuclear charge. While bound in a nucleus they are stable. They are subject to the "strong force," which keeps a nucleus intact and is one of the four fundamental forces of nature. The other three are the weak force, responsible for radioactive decay; the electromagnetic force, which governs interactions between electrically charged particles; and gravity.

When a neutron exists in a free state it decays into a proton and electron. It has a mean lifetime of about 15 minutes. Proton decay had never been observed, though some scientists began to wonder whether proton decay was possible after all. Many had rejected the notion because proton decay seemed incompatible with existing theories about proton behavior. Sakharov dared to question this orthodoxy. His 1967 letter, buttressed by new theories, became a catalyst for further research.

Sakharov postulated that immediately following the Big Bang, matter and antimatter were present in equal amounts in the new-born universe. When a particle and antiparticle meet, however, they annihilate each other. This meant that all matter and antimatter should have disappeared, leaving no massive particles at all. He suggested that the reason we live in a universe composed of matter and not antimatter could be a result of the decay of protons. His idea sparked attempts to discover evidence of proton decay. Although thus far the investigations have been unsuccessful, failure has only intensified the determination of some scientists to continue the pursuit. In 2016, the distinguished physicist Sidney Drell described Sakharov's letter as bearing an "elegant assumption" and "brilliant insight." Consequently, according to Drell, "the discovery of proton decay became ... the goal of several very-large-scale underground detectors."[1]

Whether prompted by Sakharov or other new theoretical assumptions, interest in the subject mounted. By the late 1970s, scientists from some 15 institutions worldwide were conducting or planning experiments in search

[1] "Sidney Drell on the new results related to Sakharov's proton decay hypothesis," Andrei Sakharov Foundation Forums, http://sakharovfoundation.org/forums/viewtopic.php?t=248.

of proton decay. Their methods of detection varied, as did the anticipated decay products. As far back as the 1950s, while working on neutrino detection, Fred Reines and Clyde Cowan, with Maurice Goldhaber (an eminent physicist at Brookhaven National Laboratory), had theorized that the lifetime of a proton could not be less than 10^{22} years.[2] That number, one followed by 22 zeros, is more than ten billion times the age of the universe. Scientists subsequently raised the figure to at least 10^{34} years — a number even harder to imagine.

In the early 1970s, new theories strengthened the likelihood that a proton does not live forever. But how is one to reveal, experimentally, its presumed truly minuscule chance of decay? How could evidence be gathered? Since their presumed lifetime is an average, some protons should live longer and some more briefly than 10^{34} years. Place enough zillions of them in a detector and at least one proton might decay during a few years of observation. It's like recognizing that although cars from a certain factory are known to last, on average, 200,000 miles, a few might fall apart within a few miles of the factory gate.[3]

Fred was among the surge of scientists considering experiments to demonstrate the proposition. Toward the end of the 1970s physicists from his Neutrino Group at UC Irvine joined with others at the University of Michigan and Brookhaven National Laboratory to build a large underground detector. The design was largely a creation of Lawrence Sulak, a young scientist at Michigan. It would be filled with purified water and surrounded by photomultiplier tubes. As particles traveled through the detector, they would produce distinctive cones of light. Proton decay, though extremely rare, could presumably be observable by prompting a particular pattern of light.

In searching for a location, the IMB (Irvine-Michigan-Brookhaven) group settled on the Morton Salt Mine near Cleveland. It was where Fred had established an underground laboratory 20 years earlier. In 1979, the group's proposal to the Department of Energy was approved for funding.

---

[2] F. Reines, C.L. Cowan, Jr. and M. Goldhaber, "Conservation of the Number of Nucleons," *Physical Review*, 1157, November 15, 1954.

[3] Matthew R. Francis, "Do Protons Decay?" *Symmetry*, April 22, 2015, https://www.symmetrymagazine.org/article/do-protons-decay; Ford interview, 2017–2019.

The proposal included a list of ten sponsoring scientists, most of them affiliated with Irvine. Fred was the designated IMB co-spokesman along with John Vander Velde from Michigan. The other names included five affiliated with UC Irvine: William Kropp, Henry Sobel, Jonas Schultz, John Learned, and Robert March; two affiliated with Michigan: Sulak and Daniel Sinclair; and Maurice Goldhaber from Brookhaven.[4]

With funding by the energy department and other academic sources that totaled $4 million, building the IMB detector began in 1979. The new project required excavating a cavern the size of a six-story building. Its top, bottom, and four walls were lined with plastic sheeting that sealed the space. More than 2000 photomultiplier tubes were attached to the surfaces of the enclosed space. The enclosure was then filled with 8000 tons of water and the detector began operating in 1982.

Diver Inspecting IMB Detector. Joe Stancampiano and Karl Luttrell, National Geographic Society, 1988. Reines Collection.

Two years later, a *New York Times* article was headlined, "Five Research Teams Record Possible Evidence of Proton Decay." The teams

[4]The IMB Experiment, http://www-personal.umich.edu/~jcv/imb/imb.html.

included the IMB group and others in India, Japan, Switzerland, and Utah. Fred Reines is cited as saying that 170 IMB "events" had been recorded in 200 days of operation. He acknowledged that the events could be attributed to particle interactions unrelated to proton decay. Still, reports by some of the other experimenters were, according to the *Times* article, almost "an outright claim of finding proton decay." The claim later proved to be exaggerated. The article further mentioned that recent theories had sought to place three of the four basic forces of nature — strong, weak, and electromagnetic — into a Grand Unified Theory. "One reason for the intensive and competitive search for proton decay," the article asserted, "is that it has been predicted by several so-called Grand Unification Theories."[5]

In 1985 Fred co-authored an article with John LoSecco and Daniel Sinclair, two of the Michigan-associated investigators. Their article reviewed the IMB group's findings. An unequivocal instance of proton decay had yet to be observed. Still, the authors maintained that decay was compatible with the most recent theoretical understandings.[6]

In 1983, Todd Haines began working with Fred's Neutrino Group. A native of southern California, he'd been an undergraduate physics major at UC Irvine before entering the graduate program. In 2018 Haines, now 58, shared his recollections with me at a Los Alamos coffee shop before heading to work at the nearby national laboratory. He was wearing a sports shirt, black shorts, and sneakers. His informal attire and laid-back demeanor seemed a cultural norm in Los Alamos. Haines described the city as extremely pleasant and safe. Few people lock their doors or their parked bicycles. How ironic, I thought. The birthplace of the most fearsome weapon is now ranked among the safest cities in America.[7]

Haines' career, he said, was "sort of the opposite order of Fred's." Fred had started at Los Alamos in the weapons program and later worked on neutrinos. "I started in neutrinos and now am over into the weapons physics area."

---

[5] Walter Sullivan, "Five Research Teams Record Possible Evidence of Proton Decay," *The New York Times*, January 6, 1984: 8.

[6] J.M. LoSecco, Frederick Reines, and Daniel Sinclair, *"The Search for Proton Decay,"* *Scientific American*, 252 (6) June 1985: 54–63.

[7] Los Alamos, NM Crime Rates, https://www.neighborhoodscout.com/nm/los-alamos/crime.

I asked why he wanted to make more nuclear bombs when we already had plenty. "I want to make none," he chuckled. "I don't make them. We try to understand them so we don't have to make any more." The laboratory does much more than weapons research. Its many non-military projects range from advancing medical technology to enhancing fuel efficiency. Los Alamos scientists have also collaborated in neutrino projects at Fermilab, the Sudbury Neutrino Observatory (SNO) in Canada, and elsewhere.

Haines received his PhD in 1986 for work on the IMB experiment. Fred Reines and Henry Sobel supervised his research, which calculated neutrino-induced backgrounds in the search for proton decay. His investigation helped "set lower limits" on a proton's assumed lifetime. This meant that definitive evidence for proton decay might require a larger detector or perhaps a different experimental approach.[8]

Haines recalled Fred Reines warmly:

> I had many one-on-ones with Fred in his Irvine office. He was, you know, a large man. Not large in a basketball or football player sense. But he was large in that he always had a commanding presence. He was not the kind of mentor that was, "I'm going to help you with computer codes" or things like that. Not with specifics. He was very good at imparting both a scientific wisdom and how to be a scientist.
>
> The IMB project was multi-institutional and Fred served as a spokesman. He was superb as a leader in that role. He was really fantastic. When there were issues even of a nontechnical nature, he was very good at taking the problem and working it out with other people. He would find a solution that was not just "my way" but a good way for everybody.[9]

John LoSecco was part of the IMB project from its inception and he offered a somewhat different perspective. Initially with the Michigan contingent, he remained an IMB collaborator after joining the physics faculty at Notre Dame in 1985. He was not especially impressed with

---

[8]T.J. Haines, *et al.*, "Calculation of Atmospheric Neutrino-Induced Backgrounds in a Nucleon-Decay Search," *Physical Review Letters*, October 20, 1986.

[9]Todd Haines interview, 2018.

Fred's management style, though like many others, he described Fred as creative and enthusiastic. "But sometimes," he added, "Fred's enthusiasm got the better of him." LoSecco recounted an instance when Fred was very excited because an experiment seemed to produce a particular effect. Fred quickly published the findings but apparently overlooked a statistical aberration. His report was shown to be erroneous and he was greatly embarrassed. Still, LoSecco said, he had enormous respect for Fred's abilities.[10]

The 1985 *Scientific American* article by LoSecco, Reines, and Sinclair offered an overview of the search for proton decay. LoSecco had also authored another article that turned out to be momentous. The IMB detector was developed principally to study proton decay, though its structure enabled observations of neutrinos as well. LoSecco was particularly interested in the possible detection of neutrinos emitted by supernovae. He wrote memos to the IMB group about the possibility, mentioned it at conferences, and in 1984 published an article titled, "Detectability of Supernova Neutrinos with an Existing Proton Decay Detector."[11]

A supernova is created by the explosive demise of a "dying" star. The event occurs when the star's hydrogen-based fuel runs out. Gravity quickly draws the remaining mass inward to the star's core and the resulting collapse generates huge shock waves that cause a spectacular explosion. The star suddenly may become brighter than the light emitted by an entire galaxy. The star gradually becomes dimmer during subsequent months.

Only two or three supernovae occur per century among the 100 billion-or-so stars that comprise our Milky Way galaxy. An estimated 300 supernovae also appear each year in the 30,000 named galaxies beyond the Milky Way. Most are dim because they are so far away, requiring a telescope to notice them. Once in a rare while a supernova is close enough to be visible to an earthling's naked eye.[12]

---

[10] John LoSecco interview, 2018.

[11] J.M. LoSecco, "Detectability of Supernova Neutrinos with an Existing Proton Decay Detector," *Science*, April 6, 1984: 56–58.

[12] "What Is a Supernova?" Space Place, NASA, https://spaceplace.nasa.gov/supernova/en/.

The most recent Milky Way supernova that could be seen by eye occurred in 1604, and none outside our galaxy have ever been seen without a telescope. That scarcity of visible supernovae resulted in limited interest in LoSecco's ideas. He felt frustrated: "Some people even laughed at me and said, 'You're never going to see anything'."

On February 24, 1987, while in his Notre Dame office, LoSecco received a phone call from a colleague.

> John, Do you know about the supernova?
>
> No. What supernova? When did it happen?
>
> Last night. It was noticed at an observatory in Chile. Someone there sent a telegram alerting astronomical observatories around the world.
>
> Which part of the sky?
>
> In the Magellanic Clouds.

The event had occurred in the larger of the two Magellanic Clouds, a pair of galaxies near the Milky Way. The supernova could be viewed optically only from the southern hemisphere. ("Near" is relative. The explosion occurred 168,000 years earlier and it took that many years for its light to reach our solar system.) But unlike visible light, neutrinos could pass through the Earth, including through the northern hemisphere, during their cosmic journey. Had the supernova's neutrinos been "seen" by the IMB detector? LoSecco phoned the detector's operator and learned that the previous night its power supply had partially malfunctioned. No one was onsite at 2 a.m., about the time the signal would have occurred. The tape that usually records particle events was assumed to be useless and had been placed in a pile headed for discard. LoSecco suggested that the operator retrieve the tape and spin it in the tape reader.

The result was thrilling. Ordinarily days and weeks could pass without a single neutrino reaction. The excitement in LoSecco's voice was palpable: "There had been 8 neutrinos in 5.6 seconds." At the same time Japan's Kamiokande detector had recorded 11 neutrino events.[13] The news prompted celebrations around the world. The exploded star, the first supernova to be detected that year, was designated Supernova 1987A.

---

[13] Franklyn M. Branley, *Superstar: The Supernova of 1987*, NY: Thomas Y. Crowell, 1990, http://www-sk.icrr.u-tokyo.ac.jp/sk/sk/supernova-e.html.

No one was more ecstatic about the IMB finding than Fred Reines. Ken Ford, who had encouraged Fred to join the Irvine faculty 20 years earlier, remembered that Fred "talked and talked about it." Ford laughed as he said, "It was as if God put him on this earth not just to detect neutrinos at the Savannah River reactor in 1956, but specifically to detect neutrinos from outer space."[14]

An article about Fred published by the National Academy of Sciences described the event as the birth of neutrino astronomy.

> For Reines who had long foreseen the possibility of such supernova neutrino detection, and who had pioneered the development of large water Cherenkov [light] detectors in the late 1950s, as well as other modern-day particle detection devices and techniques, it was a dream come true.[15]

Fred's passion was reflected in a commentary he wrote soon after the supernova event, titled "Stardust."

> The time: Monday the 23rd of February 1987 at 7 hours 36 minutes and 41.37 seconds (universal time). This day witnessed the occurrence of a most remarkable event — signals from inside an exploding star reached our Earth calling attention to the fiery extinction of a remote heavenly orb.... For the first time in history mankind was privileged to see deep inside a remote star. The messengers heralding this historic event were the elementary, extraordinarily penetrating particles called neutrinos. Viewed by neutrino "light" we could literally directly observe the exploding star's inner turmoil. In one step this observation transformed our understanding from hypothesis to reality; with this observation came a verification which has profound meaning for the very existence of life in the universe — including the existence of mankind.

The essay concluded with his life-long view that science could be a pathway to global civility:

> The sighting by neutrino light is an example of the universality of the scientific endeavor, achieved by the cooperation of scientists the world

---

[14] Kenneth Ford interviews, 2017–2019.

[15] Kropp, Sobel, Schultz, *Frederick Reines, 1918–1998, A Biographical Memoir, Washington, D.C.: National Academy of Sciences, 2009: 6–7.*

> around. It gives us all increased hope that this spirit of cooperation, this ability of bits of stardust to engage in rational thought, will help lead us to a world graced by a deeper harmony and common purpose.[16]

The IMB's supernova sighting was widely hailed, but the hunt for proton decay remained unavailing. Government support for IMB was fraying and in 1988, the Department of Energy sent a fateful letter to the IMB leadership. While recognizing the historic importance of the supernova sighting, the agency determined that "the usefulness of the IMB detector for proton decay will have passed the point of diminishing returns by 1989."[17]

In 1990, a massive water leakage occurred after large tears had developed in the plastic wall of the tank. Funding for repair was unavailable and IMB came to an end.

In 1988, Fred Reines became a professor emeritus. In celebration of his 70th birthday, the University of California, Irvine, hosted a daylong "Reinesfest." The event featured speakers who reviewed his numerous contributions. John Learned, a world authority astrophysicist at the University of Hawaii, had worked with Fred on numerous projects. His observations were typically adulatory. He said that Fred's "tracks are everywhere." "There is little in experimental neutrino physics that has not been touched by Fred Reines."[18]

The keynote speaker was Melvin Schwartz whose major professional accomplishment somewhat paralleled Fred's. Fred and his colleague Clyde Cowan had been the first to detect a neutrino, in 1956. As described in Chapter 7, it was at the Savannah River site that they discovered a low energy, reactor-generated, electron neutrino. Six years later, in 1962, Schwartz, Leon Lederman, and Jack Steinberger, colleagues at Columbia University, detected a second type of neutrino. It was a high-energy muon

---

[16] Reines Collection.

[17] Robert Crease, "Proton Decay Experiment on the Brink of Extinction," *The Scientist*, May 16, 1988, https://www.the-scientist.com/news/proton-decay-experiment-on-the-brink-of-extinction-62983.

[18] John Leonard, "Extra-Solar Neutrinos," in W. Kropp, M. Moe, L. Price, J. Schultz, and H. Sobel, eds., *Neutrinos and Other Matters: Selected Works of Frederick Reines*. Singapore: World Scientific, 1990: 313.

neutrino generated by a particle accelerator at Brookhaven National Laboratory in Long Island.

Leon Lederman | Melvin Schwartz | Jack Steinberger

Courtesy of the AIP Emilio Segrè Visual Archives.

Schwartz and Reines had something else in common: Each believed that his own discovery was so significant that a Nobel Prize was imminent. But every October, in the absence of a call from Stockholm, Schwartz and Reines were disappointed. By the time Schwartz spoke at the Reinesfest, he had long given up his Nobel expectation.

In 2018, Melvin's son David told me that his father, who had died in 2006, much admired Fred's accomplishments. David Schwartz, though not a physicist, had recently written a biography of Enrico Fermi, and has an impressive understanding of physics. When we met in New York City, I gave him a video of his father's keynote presentation, copied from a disc in the Reines Collection. David was thrilled to see his father on screen and he shared with me the pleasure of hearing Melvin's opening remarks:

> It's really an honor for me to speak about Fred, who has probably done more for neutrino building than anyone else. I have admired him for many, many years. He conducted one of the great classic experiments of all time. He was the first to detect, and measure the properties of, neutrinos.

Fred, sitting in the front row of the lecture hall, must have been ecstatic. Mel Schwartz, aided by charts and photographs, reviewed the Reines–Cowan experiment so that, as he put it, anyone could understand and appreciate it.

Mel Schwartz's principle contribution to his team's 1962 discovery was to configure a way to explore the weak interaction (radioactive decay) at high energy levels, using an accelerator. According to David:

> It came to him that if you created a neutrino beam you could do that. And the way you create neutrino beams is you ram protons into a target. Out come pions, which decay quickly into muons, which decay even more quickly into neutrinos. If you filter out the pions and muons you'll get neutrinos at the other end of the filter. The filter was this 30-foot thick steel wall. And it worked.

After that discovery Schwartz remained at Columbia until 1966 when he moved to Stanford University. While at Stanford, he began to develop a company called Digital Solutions that would enable people to log into their computers securely. He was torn about leaving academia to devote full time to his business. David recalled his father saying, "You know, they never give Nobel Prizes to businessmen." David's mother urged her husband to do whatever he wanted to do. Mel Schwartz left Stanford in 1983 to become a full-time entrepreneur. At that point, David said, "he really did leave behind the notion that he would ever win the Nobel Prize."

In October 1988, 5 months after Schwartz's presentation honoring Fred, he received a middle-of-the-night call from Sweden. He and Lederman and Steinberger had been named co-recipients of that year's Nobel Prize in physics. David said his father was shocked.

> When he did the experiment in 1962, in the brashness of youth he expected to win it the next year. Then, year after year he was disappointed. When he won it in 1988 he was *very* surprised. He was *really* surprised and delighted.

For Fred the announcement was devastating. He had been passed over again by the Nobel Committee for his and Cowan's 1956 discovery of the electron neutrino. The award would not be given for the first neutrino to be detected. Rather it would be for detection of a second type of neutrino, the muon neutrino. When Fred returned to work, neither he nor his colleagues discussed the disappointment they all felt.

In 1990, Fred traveled to Geneva, Switzerland, for an international conference at CERN, the European Organization for Nuclear Research. In a hall filled with more than 300 physicists and astronomers he sat next to Leon Lederman. The subject was neutrino research, and the speaker noted that two types of neutrino had thus far been discovered, the electron and muon types. A third type, associated with the tau particle, was believed to exist, though it had yet to be detected. With a straight face the speaker doubted that many investigators would try to detect the tau neutrino, "because evidently they only give a Nobel Prize for detection of every other neutrino."

The audience burst into laughter. Fred's long-time colleague William Kropp was sitting behind Reines and Lederman, the man whose team had actually received the prize. Kropp couldn't see Fred's face, but noted that unlike many around him Fred sat motionless, apparently looking straight ahead.

We cannot know what Fred was thinking at that moment but his long-held aspiration to win a Nobel surely seemed less likely to be fulfilled. That assessment paralleled another reality: An emeritus professor typically has few responsibilities and becomes less engaged in professional activity. Fred was slowing down. Still, he remained excited about science, and he was especially curious about the still-unresolved matter of whether a neutrino had mass. To answer that question two experiments were about to be undertaken in which members of his Neutrino Group at Irvine had key roles. One experiment was in Japan and the other in Canada.

Chapter 16

Super-K, SNO, and Oscillations

Two hundred miles west of Tokyo, Mount Ikeno stands 10,000 feet high in the Japanese Alps. For more than a thousand years, the mine beneath the mountain has been a source of lead, silver, and primarily zinc. In 1983, the Kamioka mine, named after the town nearby, began to yield an additional product: information about subatomic particles. Japanese physicists had built an underground detector there, about the same time that Fred Reines and the Irvine-Michigan-Brookhaven collaboration completed its detector in the Ohio salt mine. Like the IMB experiment, the Japanese project, called Kamiokande, was intended to detect proton decay.

Masatoshi Koshiba, a leading physicist at the University of Tokyo, headed Kamiokande, where he led investigations to detect cosmic neutrinos. Koshiba's findings also confirmed Raymond Davis' identification of solar neutrinos, for which they jointly received the 2002 Nobel Prize in physics.

Masatoshi Koshiba Raymond Davis
Courtesy of the AIP Emilio Segrè Visual Archives.

The Kamiokande detector, like the IMB detector, consisted of a large tank of purified water. The cylindrical Kamiokande tank, 50 feet tall and 50 feet in diameter, contained 3000 tons of water. It was lined with 2000 photomultipliers and its water presumably provided an ample supply of protons (hydrogen nuclei) to prompt signals of proton decay.

Takaaki Kajita, a 24-year old graduate student at the University of Tokyo, was among the young scientists brought on by Koshiba to participate in the project. In 1986, three years after the experiment's inception, Kajita received his PhD for work on proton decay. That same year Todd Haines had gotten his PhD from Irvine for a similar effort with the IMB detector. Haines described the IMB and Kamiokande groups as "pretty competitive." Both Haines and Kajita had produced useful information, but neither of them, nor anyone else, had observed proton decay.

While engaged in the effort, Kajita became interested in a matter that harked back to Fred Reines' discovery two decades earlier. Reines' experiment in the South African Rand gold mine had detected atmospheric neutrinos. They were created by cosmic ray interactions 5 or 10 miles above the Earth's surface and Kajita was curious about the relative proportion of electron neutrinos and muon neutrinos that were passing through the Kamiokande detector. He assessed the numbers of each type — or flavor — of atmospheric neutrinos that were being recorded in the detector. He described the result as "strange." The number of muon neutrino events was much lower than expected. "At first I thought that I made a serious mistake," he said.[1]

He reexamined his data and procedures but could not find any errors. He and some colleagues wondered whether a software program for the experiment might have been faulty. Their continuing assessment was interrupted in February 1987 by the sudden appearance of Supernova 1987A. Just as neutrinos emitted by the supernova were belatedly recognized on the IMB detector's tape, the Kamiokande discovery was also almost missed. Stephen Brice, head of the Neutrino Division at Fermilab in Illinois, recounted what he'd learned about the experience:

> They had shut down the Kamiokande detector for some maintenance reasons an hour before the neutrino flux from Supernova 1987A arrived. They only brought it back up maybe half an hour before, because someone had decided, Oh there's one more thing we have to do. Or have you double checked that? Or whatever. It could have been completely missed.[2]

Kamiokande's detection of 11 events and IMB's of eight events in a period of a few seconds became international news. The findings demonstrated what until then had been just a presumption, that a supernova emits a great many neutrinos as well as a lot of light. Brice further noted:

> We haven't detected a supernova by way of neutrinos since then. We're constantly on watch. We've detected supernovae in light, multiple times in a year. They all produce vast quantities of neutrinos as well as

---

[1] Kajita's remarks are drawn from his 2015 Nobel address and subsequent lectures at UC Berkeley https://events.berkeley.edu/?event_ID=111294&date=2017-11-01&tab=all_events; and at the University of Connecticut, https://physics.uconn.edu/home-2/events-special/.

[2] Stephen (Steve) Brice interview, 2018.

light. But mostly they're too far away, so that by the time the neutrinos come to us the flux is so small that we're not likely to see it. So we're only likely to see in neutrinos a supernova that occurs in our own galaxy. Whereas in light we can see supernovae in many of the galaxies around us.

After excitement about the supernova experience subsided, the Kamiokande scientists returned to pondering possible causes of the muon neutrino deficit. In 1988, after a year of further study, they published their puzzling results. The 85 muon neutrino events that they had observed were little more than half the 144 that had been predicted. Their paper concluded:

> We are unable to explain the data as the result of systematic detector effects or uncertainties in the atmospheric neutrino fluxes. *Some as-yet-unaccounted-for physics such as neutrino oscillations might explain the data.*[3]

Oscillations would mean the spontaneous changing back and forth of neutrinos from one type to another. Kajita added a personal note 30 years later, in 2017, when citing the article's passage during a presentation at UC Berkeley. He recalled being "most excited" about the possibility of neutrino oscillations. It gave "me a strong motivation to continue the study."[4]

In his Berkeley lecture, Kajita said he had come to think that the reason oscillations had not been observed might be related to the distance they traveled. Atmospheric neutrinos close to the Earth traveled downward into the detector. Perhaps there was insufficient time for them to oscillate. This would not be the case for atmospheric neutrinos coming from the other side of the Earth. Their trip of up to 8000 miles through the planet could allow time for oscillation. Also, these neutrinos could be identified by their trajectory. They would be entering the detector from below and traveling upward through it.

---

[3] K. Hirata, *et al.*, *Physics Letters B*, 205 (1988): 416.

[4] Takaaki Kajita lecture, 2017. https://physics.berkeley.edu/news-events/events/20171030/2017-emilio-segre-lecture-featuring-takaaki-kajita.

Kajita and others understood that an experiment to test the oscillation hypothesis would require a much larger detector than Kamiokande. In 1991, the Japanese government agreed to fund the proposed Super-Kamiokande to look for nucleon decay. The new detector would also advance capabilities for neutrino research. At a height of 130 feet and a diameter of 130 feet, its cylindrical frame would be nearly 20 times the size of the predecessor Kamiokande. The new structure would contain 50,000 tons of water.

Meanwhile, the IMB project had ceased to exist and Kajita invited IMB scientists to become part of the Super-K experiment. During his 2017 lecture at Berkeley, Kajita showed a photograph labelled "Beginning of the Super-Kamiokande collaboration between Japan and USA." The collaboration began in 1991. In the picture, Kajita is standing in the middle of a group of 10 men, seven of them Japanese and the three others Americans who had been part of IMB. James Stone, though on the faculty of Boston University, was associated with the Michigan contingent. Henry Sobel and William Kropp were Irvine researchers, both of them close colleagues of Fred Reines. Their careers had begun as Fred's graduate students at Case Institute a quarter-century earlier. Fred was unable to participate actively in Super-K, but Sobel and Kropp kept him informed about its progress.

Beginning of Japanese and U.S. Super-Kamiokande Collaboration. Courtesy of Takaaki Kajita.

The Americans' trip to Super-K began with a 12-hour flight from Los Angeles to Tokyo followed by a one-hour air shuttle to the west coast city of Toyama. From there a 90-minute drive inland took them to the Japanese Alps. The narrow road curled up and around rocky crags. A lapse in a driver's concentration could result in a catastrophic spill into a steep ravine. Still, travelers were in the midst of scenic beauty. In winter, snow-covered peaks jutted upward from dark rolling plains. In spring the plains turned green, a marked contrast to the surrounding rocky heights. Every season summoned a new and splendid vista. The American crew was headed for Mount Ikeno.

By 1994 the Americans had become frequent travelers to Super-K. Their visits to Japan sometimes lasted for months as they helped construct the new detector. The Super-K detector was divided into two sections that would eventually be filled with pure water. The first section was the huge cylindrical tank whose inner surface was covered with 11,000 photomultiplier tubes. The second section was a narrow sealed-off space that surrounded the inner tank. The outer detector, also lined with phototubes, could distinguish neutrino interactions from cosmic ray muon events. The outer section was known as the "veto" detector. Kropp explained:

> Something coming in from outside goes through the outer and then the inner detector. The outer detector is the veto detector because if you see an object like a cosmic ray there as well as in the inner detector you "throw it away," because that tells you it is not a neutrino. But if you see it only in the inner detector, and not anything is happening in the outer detector, it could be a neutrino.[5]

The distinction was possible because of the difference in light patterns that the particles emitted when entering the detector. Unlike the neutrino, which has no electric charge, the electrically charged muon emitted a light signal upon entering the outer detector. As described on the Super-K website:

> Because the neutrinos are electrically neutral, the Cherenkov light is emitted when the neutrinos scatter with the water. On the other hand, because the cosmic ray muons have an electric charge, the Cherenkov

[5] Kropp interviews, 2017–2019.

light is immediately emitted when the cosmic ray muon enters the detector.... This difference enables [distinguishing] between neutrinos and cosmic ray muon events."[6]

An initial responsibility for the Americans was to detach the 1800 photomultiplier tubes from the IMB detector and ship them and related items to Japan. Once there, under Kropp's supervision, the Americans constructed the outer detector whose distance between walls measured only a few feet. Unpurified water was poured into both detectors to reach varied heights. In the narrowly spaced outer detector, individuals stood on floating platforms to attach the phototubes to the steel scaffolding that lined the walls. Among the American contributors to the effort were Lawrence Sulak, now at Boston University, and other former IMB scientists. Dozens of graduate students from Irvine, Boston, and other American schools were also recruited for brief periods to help install the tubes.

Japanese personnel built the inner tank. In the cavernous inner detector, scientists and other workers transported phototubes on a small rubber boat. They would row to a designated location along the wall. After attaching a tube to the steel scaffolding, they would move around the circular wall installing tubes until the row was completed. The technique was far safer than using ladders or elevators, Kropp observed. "Anyone who fell would only fall into the water, not a hundred feet through open space."

In 1996 construction was completed. The impure water that carried the boats and the floating platforms was drained out and the entire detector was filled with ultra-clean water. The electronics were turned on, and the experiment was underway. During the next two years, Super-K was able to identify about 10 neutrino events per day. They were both electron-neutrino and muon-neutrino events. At a neutrino conference in 1998, Kajita reported that the data "showed neutrinos coming from above and neutrinos coming from the other side of the Earth. And you can see there is a huge deficit of muon-neutrinos coming from the other side of the Earth. [We] concluded that the data gave evidence for neutrino oscillations."

---

[6] Super-Kamiokande Detector http://www-sk.icrr.u-tokyo.ac.jp/sk/detector/index-e.html.

Takaaki Kajita
Courtesy of Wikimedia Commons.

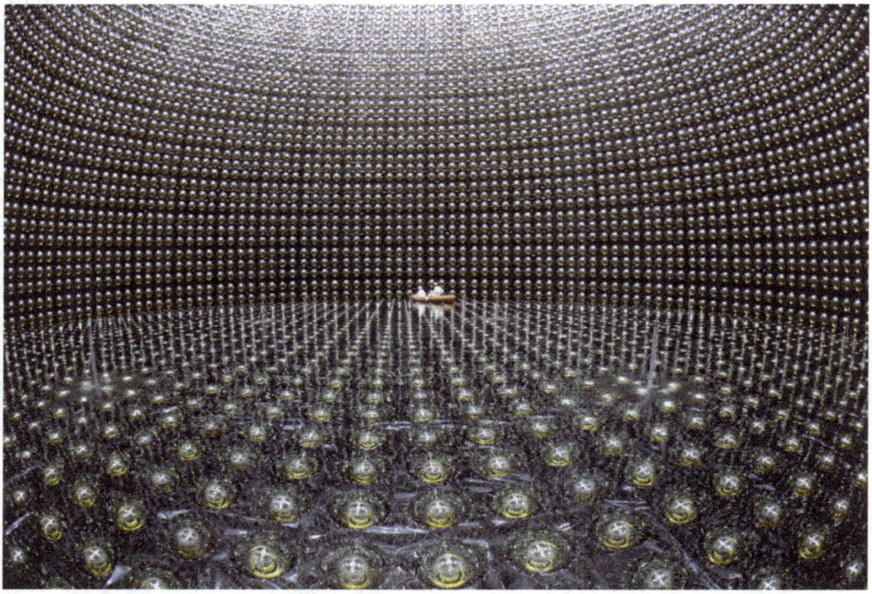

Super-Kamiokande Neutrino Detector. Courtesy of the Kamioka Observatory, ICRR (Institute for Cosmic Ray Research), The University of Tokyo.

The Super-K's demonstration of oscillations was followed a few years later by complementary evidence from the SNO (Sudbury Neutrino Observatory) experiment in Canada. Members of the Reines Neutrino Group at Irvine were key figures in both experiments. Sobel and Kropp were founding members of the Super-K collaboration and Herbert Chen was central to the development of SNO, until his untimely death in 1987. These scientists and other participants from the Irvine group had been mentored, inspired, and deeply influenced by Fred.

Prior to Chen's interest in solar neutrino studies, he conducted preliminary investigations with liquid argon as a detection medium. At the time he was at the Los Alamos National Laboratory, on leave from Irvine. In November 1977, he received a letter from a British student who had heard of his work. Peter Doe was about to receive a PhD from Durham University. He too was interested in liquid argon as a particle detector and wondered if Chen might consider hiring him. After sending the letter, Doe received a telephone call from Chen who simply asked, "When can you be here?"

Doe, then a designated member of the Irvine physics department, arrived at Los Alamos two months later. He joined Chen on some small argon detector projects, and later worked with him when Chen began to seek a more expansive base for inquiry. As Doe recalled:

> Herb was trying to find a location to build a big underground detector using liquid argon, which would detect solar neutrinos. At that time there was the solar neutrino problem. He was talking to some Canadian colleagues in Toronto. One of the locations they were thinking about for a large detector was the Creighton nickel mine in Sudbury, Ontario.[7]

In the early 1980s, Chen realized that heavy water might be preferable to argon as a detection medium. Until then, solar neutrino detection methods had been sensitive to electron neutrinos but not to the other flavors — muon or tau neutrinos. Years earlier Fred Reines had surmised that heavy water could be an effective detector of solar neutrinos. Now

---

[7] Peter Doe interview, 2018.

Chen understood that heavy water could also be used to detect all the neutrino flavors.

The hydrogen nucleus in ordinary water is a proton. The water is "heavy" if nearly all of its hydrogen nuclei consist of one proton and one neutron. On average one heavy water molecule exists naturally in about 7000 molecules of normal water, and producing concentrated volumes of heavy water is difficult and expensive. Although sometimes used in small projects, the cost and limited availability of heavy water had kept it out of reach for large experiments.

Canadian nuclear reactors, unlike many reactors elsewhere, use heavy water as a coolant. Chen mentioned to Canadian colleagues that the liquid would be preferable to argon for his experiment. The Canadians agreed that heavy water would be an optimal detection medium for a neutrino experiment. Doe recalled the conversation:

> Heavy water is expensive, about $1,000 per liter. Herb said we can't afford this. A Canadian guy, Walter Benjamin, said they have a lot of this stuff laying around, because it was part of the Canadian reactor program.

The required volume would cost about $300 million. When Benjamin conferred with government colleagues they agreed that 1000 tons of deuterium oxide — heavy water — could be lent to the project and later returned to the government. Upon hearing this, Chen dropped the idea of using liquid argon and accepted the Canadian offer. He also agreed that the laboratory be built in the Sudbury nickel mine.

In 1984, 16 scientists, mainly from Canada and the United States, began collaborating on the SNO project. At the outset, Chen was designated leader of the American team and George Ewan of the Canadians.[8] In 1986 Chen was diagnosed with leukemia and, tragically, the following year he died at age 43. At that time, Arthur McDonald, a Canadian professor at Princeton, became the U.S. spokesman for the project. According to McDonald:

---

[8] G.T. Ewan, W.F. Davidson, "Early Development of the Underground SNO Laboratory in Canada," *Physics in Canada*, Nov.–Dec. 2005: 340.

> Fred Reines was very helpful in maintaining the momentum of the project at a difficult time for us, both through temporary preservation of the funding at Irvine and finding new positions for Herb's postdoctoral fellows, and also through advice on additional U.S. collaborators.[9]

In 1988, McDonald spent a sabbatical year in Canada, working on the final project design and funding the proposal. He subsequently accepted a faculty position at Queen's University in Kingston, Ontario, and became the overall Director of SNO. Years later he recalled that "Chen's ideas and his drive had gotten the project going, and continued to inspire us after his death."[10]

When Chen died, Fred Reines, like many others, was crestfallen. Two months after Chen's death, Fred moderated a memorial symposium at UC Irvine on neutrino physics. Physicists delivered scientific presentations throughout the day. Fred also co-authored an obituary that included the following:

> It required courage to undertake the notoriously difficult experiments that became Chen's hallmark, and he faced his illness with the same courage. Although frequently hospitalized after the onset of his illness, he continued to guide and advise from his hospital bed via modem and telephone, promoting the physics he loved. In a tragically short career he achieved much; the seeds of ideas he left behind promise even more.[11]

In 2001, the SNO project succeeded in determining that the electron neutrinos created in the sun's core could oscillate between electron, muon, and tau flavors as they travel through space. The oscillations are analogous to a mixture of three ice cream flavors. Each flavor has a distinctive mass and wave, and as a neutrino travels, the waves mingle and advance at different rates. At different points the mixture of flavors may differ — sometimes the chocolate taste would dominate, and sometimes vanilla or strawberry. Thus, the understanding that "neutrinos could oscillate

---

[9] Arthur (Art) McDonald, email communication, 2019.

[10] McDonald, Annual Reines Lecture Series, UC Irvine, March 2017.

[11] Richard Allen, Peter Doe, and Frederick Reines, "Herbert H. Chen," *Physics Today*, September 1988: 128, https://physicstoday.scitation.org/doi/pdf/10.1063/1.2811575.

between flavors as they travel from the sun to the Earth."[12] With this determination, the SNO experiment confirmed that the "missing" neutrinos had been found.

In 2015, McDonald was a co-recipient of the Nobel Prize for physics with Japan's Takaaki Kajita. The platform for McDonald's work at SNO had been initiated by Chen. It is commonly believed that had Chen lived, he would have been a third Nobel co-recipient.

In 2017, two years after receiving the Nobel Prize, McDonald delivered the 3rd Annual Lecture in a series at UC Irvine honoring Fred Reines.

Arthur McDonald. Courtesy of the AIP Emilio Segrè Visual Archives.

McDonald's lecture included reference to the difficulty in constructing the huge spherical detector vessel:

> The acrylic vessel had to be cut into 120 pieces. Once down in the excavated area for the experiment, the acrylic pieces had to be carefully positioned on a jig and bonded together. It took over 2 years to do that.

[12] Jayawardhana: 98–99.

Once reconstructed, the vessel was filled with heavy water and surrounded by a geodesic sphere.

> [The outer sphere] contains 10,000 light sensors that cover almost all the surface area of the detector. That concentration of light power would be adequate to see a candle on the moon.[13]

SNO Neutrino Detector (Arthur McDonald, Reines Lecture Series, UC Irvine, 2017). Courtesy of SNO Collaboration and Lawrence Berkeley National Laboratory.

With the use of heavy water as a detection medium, the previously unseen muon and tau forms were now included in the total flux of neutrinos that were reaching the Earth. A scientific paper that described the SNO finding was co-authored by 230 collaborators. Among the listed Irvine contributors were Hebert Chen, Peter Doe, and Richard Allen. In addition, other articles on SNO referenced earlier work by other members

---

[13] McDonald, Annual Reines Lecture Series, UC Irvine, March 2017. Also see: "How to Know Neutrinos — with Art McDonald," https://www.youtube.com/watch?v=eWc0jywR-H4.

of the Irvine Neutrino Group including William Kropp and Henry Sobel, and of course, Fred Reines.[14]

McDonald later recalled that after Chen's death, Fred became the active interim leader of the Irvine group and continued to provide advice on the project. McDonald also said that he "always respected Fred as a visionary experimental physicist and gained greater respect [for him] with the way he provided assistance for us at a difficult time."[15]

In 1998, the Super-K detector had found apparent evidence that neutrinos oscillated and therefore had mass. Three years later the SNO experiment in Canada directly observed neutrino oscillations, confirming the non-zero-mass finding by the Super-K. Members of the Irvine Neutrino Group had been essential collaborators in both projects.

At the conclusion of McDonald's 2017 Reines Lecture, he raised fundamental issues of the sort that Fred had addressed throughout his career:

> You might ask, Why do you care? Why should you want to do these measurements about the sun and neutrinos? How is it going to affect me in any way? It turns out that most of the elements from which we are made, including very fundamental elements like carbon, oxygen, and nitrogen are made in nuclear reactions that occur in the cores of stars like our sun.

Earlier in his lecture McDonald provided a more encompassing answer to the question, Why study neutrinos?

> You're really studying the whole range of the universe in the sense that you're studying the most microscopic particles that we know, in the form of neutrinos, and potentially also dark matter particles. You're studying the sun and how it burns. You're studying the Earth by looking at neutrinos coming from the Earth. You're studying the farthest reaches of

---

[14] Reines Collection; Q.R. Ahmad, *et al.*, "Direct Evidence for Neutrino Flavor Transformation from Neutral-Current Interactions in the Sudbury Neutrino Observatory," *Physical Review Letters*, June 13, 2002, https://journals.aps.org/prl/abstract/10.1103/PhysRevLett.89.011301.

[15] McDonald, email communication, 2019.

the universe in the sense of studying things that relate to the origin of the universe.

McDonald then recited a series of issues that he believed the study of neutrinos could help resolve: how the elements from which we are made are created in the sun and other stars like it; what the basic laws of physics are for the smallest fundamental particles; what the overall composition of our universe is; and how the universe has evolved to the present time.

Chapter 17

The Nobel Prize and the Last Years

By the mid-1990s, Fred was largely disengaged from academic and scientific activity. He had ceased both teaching classes and supervising graduate students. Lawrence Sulak, an IMB colleague who later collaborated on Super-Kamiokande, recalled that Fred strongly advocated building an outer veto area in the new detector. After construction began in 1991, however, Fred was no longer an active participant. Still, he periodically went to his office, conferred with physics colleagues, and shared thoughts with friends.

Early in 1995 LeRoy Price, a young colleague in the Neutrino Group, spoke with Fred about his long-held though unfulfilled expectation that he would receive a Nobel Prize. In 2018 Price recounted the conversation to me:

> Fred told me that he had finally given up hope of getting the Nobel Prize. And he meant it. "I know there are many people who should have gotten the prize and didn't get it," he said. "Maybe some got it that didn't deserve it. It's not up to me. But it's clear that since I've not gotten it by now that I'm not going to get it." He spoke not exactly with sadness, but more or less with resignation. You know just, "OK, I give up." What was mind-boggling to me was that was the year he got it.[1]

The first Nobel Prize in physics was awarded in 1901 to Wilhelm Roentgen for his discovery of X-rays. Four years later Albert Einstein, an

[1] Price interview, 2018.

obscure 26-year-old German-born Swiss patent clerk, published four scientific papers. The first was on the photoelectric effect — the ejection of electrons from metal surfaces that are irradiated by light. Einstein accounted for all aspects of this phenomenon by postulating that light consists of discrete energy packets, later named photons. His second paper in that "miracle year" was on Brownian motion — the jiggling of small particles suspended in a liquid (or a gas). His achievement here was to show how the phenomenon could reveal the sizes of atoms. The third paper introduced what we now call special relativity, based on the proposition that the speed of light is constant for all observers. The fourth, on the equivalence of mass and energy, gave rise to his most famous equation, $E = mc^2$. A decade later came Einstein's paper on general relativity, which further altered traditional conceptions of space and time and revolutionized our understanding of the universe. No wonder that Einstein is recognized as one of the greatest scientists that ever lived.

Any one of Einstein's contributions to human understanding seems now to have been worthy of a Nobel Prize soon after its discovery. Early recognition by the Nobel committee would have been consistent with the wishes of the award's benefactor. Alfred Nobel, a Swedish millionaire who invented dynamite, died in 1896. He left $9 million to fund annual prizes in Physics, Chemistry, Physiology or Medicine, Literature, and Peace. (A Memorial Prize in Economics was added in 1968.) Nobel's will specified that the prize go "to those who, during the preceding year, shall have conferred the greatest benefit on mankind." The timeliness directive has repeatedly been ignored.

Eminent scientists recommended or formally nominated Einstein for the prize almost every year after 1909. Einstein biographer Abraham Pais cited dozens of letters to the Nobel Committee with wording such as, "the Nobel Committee will rarely have the opportunity of awarding a Prize for works of similar significance." Or, Einstein was "in the first rank of physicists of all time." Perhaps the most trenchant letter came from French physicist Leon Brillouin: "Imagine for a moment what the general opinion will be fifty years from now if Einstein's name does not appear on the list of Nobel laureates."[2]

---

[2] Abraham Pais, *Einstein Lived Here*. NY: Oxford University Press, 1994: 70–74.

Einstein finally received the prize in 1922, but only for his explanation of the photoelectric effect. (Ironically, his proposal that light consisted of particles was not generally accepted until after he received the prize and evidence was then provided by others.) The award came late for a tangle of reasons. They included the Nobel committee's apparent preference for physics experimenters over theorists, petty politics and jealousy, anti-Semitism, and lingering skepticism by some about the validity of relativity.[3]

In fact, Einstein's receipt of the award may have added more luster to the Nobel Prize than to Einstein's reputation, which in 1922 was already spectacular. In the years following, the name "Einstein" commonly became synonymous with unparalleled brilliance. At the end of 1999, *Time* magazine named him first among the 100 most important people of the 20th century.[4]

For many, a Nobel Prize has become a symbol of virtual deification. Brian Keating, a professor of physics and astronomy at the University of California, San Diego, recounted the time Duncan Haldane displayed his 2016 Nobel gold medal after a lecture.

> A throng of physicists young and old got caught up in the excitement. They fawned over the gilded medallion, pushing past one another just to get a glimpse of it. No one actually bowed down to it. But some did kiss it. Some tried to sneak off with it. I'm ashamed to say I was among the worshippers, unable to resist posing for a picture with the medallion.[5]

Keating applauds the original purpose of the prize: to promote an inspirational and meritocratic reward system. He believes, however, that the current criteria for selecting winners have warped Alfred Nobel's original intentions. He thinks the Nobel Committee has inappropriate authority in umpiring a scientist's chase toward discovery. The Nobel Prize has become, according to Keating, "a reward mechanism that

---

[3] Burton Feldman, *The Nobel Prize: A History of Genius, Controversy, and Prestige*. NY: Arcade, 2000: 145–148.

[4] Frederic Golden, "Albert Einstein," *Time*, Dec. 31, 1999.

[5] Brian Keating, *Losing the Nobel Prize, A Story of Cosmology, Ambition, and the Perils of Science's Highest Honor*. NY: W.W. Norton, 2018: 267–268.

discourages collaboration, celebrates authority, and sets up a rat race where claim-staking, speed, and greed are encouraged."[6]

Whatever one thinks of Keating's assessment, in Fred's case the speed with which he and Cowan published their findings was irrelevant. Nearly forty years passed before the Nobel judges recognized the Reines–Cowan achievement. Fred's middle-of-the-night call from Stockholm came on October 10, 1995. Sylvia took the call because Fred was in the hospital. His vitality had diminished substantially, and he was there for observation and care.

Fred's daughter, Lisa Cowden, and her family were living in upper New York State. At around 7 a.m. she yelled from upstairs down to her daughter Julia to quickly turn on the NPR radio station. Julia, a high-school junior, recalled feeling awed. She later wrote about her experience: "It was exciting to hear my grandfather's name on the radio, even though I didn't understand fully what getting a Nobel Prize meant."[7]

After the announcement several colleagues visited Fred in the hospital, expressed their excitement, and offered to help in any way they could. In the following days, Fred became more alert as the full realization of the news sank in. He accepted his colleagues' offers for help in preparing the lecture that is expected of each new laureate.

The Reines family arrived in Stockholm in advance of December 8th, for a week of celebratory events: banquets, entertainment, panel discussions with the new laureates, interviews, speeches, press conferences, and ultimately King Gustav Olaf's presentations of the gold Nobel medals. Dress for the presentation ceremony was entirely formal, men in tails and women in evening gowns. The ceremony began with stirring music from the Royal Stockholm Philharmonic. The laureates were hailed with the sound of drums and trumpets. Julia described the proceedings:

> The president of the Nobel Foundation gave a speech in Swedish. Then a citation for each prize was read in Swedish and when the laureates' names were called, they stood up and walked forward to meet the king, who gave them their prize. Due to the friendship between Martin Perl, the other laureate in physics, and Grandpa, they went up to receive their

---

[6] *Ibid*: 263.

[7] Julia Cowden, "Nobel House, A Trumansburg Girl's Trip to Sweden," Reines Collection.

prizes together. There wasn't a dry eye in the hall, because my grandfather had been ill and Martin Perl took Grandpa's arm and walked with him to the center of the stage.[8]

Receiving the Nobel Prize in Physics from the King of Sweden, 1995. Courtesy of Associated Press Photographs.

Coincidentally one of Fred's colleagues at UC Irvine also became a Nobel laureate in 1995. F. Sherwood Rowland was a co-recipient of the prize in chemistry. It was a year of double celebration for the university and especially for two of its physical science departments, which Fred had overseen as Dean.

Visibly frail when he received the Nobel Prize in 1995, Fred walked and talked slowly. Still, he and his family savored the experience. They were, as his daughter Lisa said, treated royally by the Nobel staff.

---

[8] *Ibid.*

Sylvia and Fred Reines at the Nobel ceremony, 1995. Courtesy of The Los Alamos Monitor.

Fred's Nobel Lecture included not only a review of his major scientific accomplishments, but also his appreciation of an open environment. After describing his and Cowan's 1956 experiment to detect neutrinos, he made note of the conditions that helped their efforts.

> But of one thing I am certain: the open, free communication of our ideas was most stimulating to us and played a significant role in our eventual success. We were not inhibited in our communication by the concern that someone would scoop us. Neutrino detection was not a popular activity in 1952.

His lecture included reference to the possibility of neutrino oscillations, which at that time was still speculative. He referred to earlier conjectures by Bruno Pontecorvo and others —

> that the neutrino may oscillate from one flavor to another as it travels from its place of origin. A graphic analogy is the change of character from dog to cat: Imagine at time zero a dog leaving his house to walk down the street to another dog house at the end of the block. As he progresses down the street a transformation takes place — his appearance gradually changes (à la Escher) from that of a dog to that of a cat! Halfway down the block the transformation is complete and the erstwhile dog — now a cat — continues on its feline journey. But the transformation goes on and, *mirabile dictu*, upon arrival at the dog house the erstwhile dog turned cat is once again a dog. If such bizarre behavior is observed to occur in neutrinos it would provide evidence of the neutrino's structure. Neutrinos of all types would be construed to be built out of common building blocks whose rearrangements en route would give rise to observably different combinations.

Fred's lecture concluded with an acknowledgement of the support he received from many coworkers:

> One in particular I must mention — my very good friend and colleague Clyde Cowan, who was an equal partner in the experiments to discover the neutrino. I regret that he did not live long enough to share in this honor with me.

Later in the week all the laureates participated on a panel titled "Nobel Minds: Challenges of the 21st Century." The televised discussion is preserved on a disc in the UC Irvine's library Reines Collection. The seven Nobelists sat in a half circle before a large audience. At one point the discussion leader, CNN newscaster Jonathan Mann, said that Fred had previously mentioned to him that secrecy could be a problem. Mann wondered how secrecy might be an issue in the realm of technology or science or politics. "Dr. Reines," he said, "What is your concern?" Fred responded:

> That there be too much secrecy. I would think the world would be a better place if A knew what B was doing, and B knew what A was up to. Openness, I think, should be the goal. In some sense there has been a

> breakdown of secrecy with the ending of the Cold War, and that's a positive thing. At some point, if you work at it and if you're lucky, you might recognize that the free exchange of knowledge is a way to pursue a better future.[9]

Other laureates responded. One suggested that industrial secrecy could be a legitimate tool in developing new products. Another said that a scientist might withhold information about an experiment until certain that the results were valid. Martin Perl, who had discovered the tau particle, shared the physics prize with Fred. He asked if Fred was worried about people keeping things secret in science. Fred acknowledged that scientists might think there are good reasons for keeping secrets. After a dramatic pause, as if about to offer a profound observation, he continued, "such as wanting to be first." The droll remark prompted laughter from the audience.

Fred's answer was a reminder of his dry wit, often expressed in the form of an unexpected word or idea. Years earlier, during a lecture to an audience of non-scientists, he referred to light that "bounces off in some kittywampus directions." Kittywampus? Perhaps some wordsmiths understood that the term meant "askew." Others may have thought it an obscure physics term until Fred's further explanation prompted sighs of relief. A few years after I had completed training in dental medicine, Fred asked, "So, how goes the tooth game?" He was as amused by his verbal coinage as I was. It was on a par with a later reference by another physicist to the sad state of current politics as "truth decay."[10]

Although Fred's vivacity had been ebbing, his habit of making whimsical comments had not. In 1996, a year after the Nobel ceremony, Fred attended a physics conference in Finland. His long-time friend John Bahcall approached him. Bahcall's wife Neta, an astrophysicist at Princeton University, was a key figure in work related to the Hubble telescope. Aware of a buzz that Fred might be having memory challenges, Bahcall asked if Fred knew who he was. Fred's playful answer: "Yes, you're the husband of the famous astronomer, Neta Bahcall."

Meanwhile, Fred was invited to receive an honorary doctorate at New York University's 1996 commencement. He had received his PhD in

---

[9] Reines Collection.

[10] Rush Holt, "Democracy's Plight," *Science*, February 1, 2019, http://science.sciencemag.org/content/363/6426/433.

physics from NYU 56 years earlier. Doubtless the Nobel Prize inspired the school to join in celebrating his achievements. By chance, in 2018 I met George Daly, who had been Dean of the NYU Business School during the 1990s. Our conversation went something like this:

> George, did you attend the NYU commencement in 1996?
>
> I attended all of them when I was dean.
>
> Do you happen to remember that Fred Reines received an honorary degree that year?
>
> Was he the scientist?
>
> He had just won the Nobel Prize in Physics.

Daly hadn't remembered Fred's name, but said he was especially impressed by him at a pre-ceremony dinner at which each 1996 honoree spoke. The others included well-known figures, most of whose comments seemed perfunctory. In contrast, Daly remembered Fred's remarks about the importance of science as deeply felt and convincing. The other honorary degree recipients included the actor Robert DeNiro; King Hussein of Jordan; screen writer Steven Spielberg; Rachel Robinson, the widow of Jackie Robinson; and Wanda Ferragamo Miletti, president of the Ferragamo luxury goods empire.

Receiving honorary degree from New York University, 1996. Reines Collection.

Fred's accomplishments were also celebrated elsewhere, including at Los Alamos and at UC Irvine. In May 1997, the university renamed its Physical Sciences II building Frederick Reines Hall. A dedication ceremony was held in the building's courtyard where a bronze bust of Reines was unveiled and then placed in the hall's foyer.[11]

UCI News

For the campus community

www.communications.uci.edu/~inform/

Reines Hall debuts

The career of UCI professor emeritus Frederick Reines, whose scientific research confirming the existence of neutrinos earned him the Nobel Prize in Physics, was honored May 12 when university officials renamed Physical Sciences II as Frederick Reines Hall. In addition, a bust of Reines created for display in the foyer was unveiled.

"Dedicating this building in his honor serves as a symbol of our respect and admiration of his achievements," Chancellor Laurel L. Wilkening said, "and of our appreciation for his long-standing devotion to UCI and its students."

Reines came to UCI in 1966 as the university's founding dean of physical sciences, a position he held until 1974.

In 1995, Reines won the Nobel Prize for research that led to confirmation of the existence of the neutrino, an abundant but elusive subatomic particle.

Reines' Nobel highlighted a scientific career marked with national and international awards and prominent research posts. He was named a Guggenheim Fellow and an Alfred P. Sloan Fellow, for example, became a member of the National Academy of Sciences in 1980 and a year later received the J. Robert Oppenheimer Prize.

Reines was honored with the UCI Medal for Outstanding Research in 1985. During that same year, President Ronald Reagan awarded Reines the National Medal of Science.

It's a bust!
A bronze sculpture of Frederick Reines was unveiled this month as part of a ceremonial renaming of Physical Sciences II in honor of the Nobel Laureate. The bust was created by artist Garth Tietjen.

Physical Sciences II renamed Frederick Reines Hall, 1997.

In 1998, Fred's health worsened and he became bedridden. Toward the end of summer, a visiting colleague told him that Super-Kamiokande leaders had announced they had evidence that neutrinos oscillated. Fred, with faint voice, asked, "What's the mass?" (The mass states of the three

[11] "UC Irvine Will Name Building for Nobel Prize Winner," *Los Angeles Times*, May 9, 1997, http://articles.latimes.com/1997-05-09/local/me-57156_1_nobel-prize.

Frederick Reines Hall.

flavors of the neutrino are still unknown.) Weeks later, on August 26th, he passed away.

Excerpts from three obituaries collectively highlight the varied corners of Fred Reines' persona and accomplishments:

From *The New York Times*:

> Dr. Frederick Reines, whose fine baritone might have landed him in opera instead of the career in neutrino physics that culminated in the 1995 Nobel Prize, died on Wednesday night at the University of California-Irvine Medical Center in Orange, Calif. Dr. Reines was 80 and lived on the university campus, where he was distinguished professor emeritus in physics.
>
> The university said he died after a long illness.
>
> Described by other scientists as the father of neutrino physics, Dr. Reines had a reputation as an intense researcher, often ahead of others

in anticipating important physics problems and always relentless in solving them.

This was how Dr. Reines and a colleague made the discovery that eventually earned him the Nobel accolade.

In the early 1930s, physicists predicted the existence of neutrinos, virtually massless subatomic particles that move through space and matter with so few interactions that their existence was thought to be beyond detection. The theorists who predicted that such ghostlike particles probably originate in nuclear reactions in the Sun and in exploding stars despaired of ever discovering their traces.

In 1951, while at Los Alamos Scientific Laboratory in New Mexico, Dr. Reines decided to try the seemingly impossible: detect a neutrino. It was typical of him. "He was always doing courageous things, expanding the limits of research," said Dr. Jonas Schultz, another physicist at the Irvine campus. "He thought of ways to detect neutrinos, showed that it could be done and that was a psychologically important breakthrough in the whole field."

Working with Dr. Clyde Cowan, a Los Alamos colleague, Dr. Reines used a nuclear reactor to produce neutrinos and a tank of water as the agent of detection. The impacts of neutrinos coming from the reactor produced flashes of light in the water tanks, which the two scientists recorded and reported in 1956. Scientists were surprised that it took the Nobel Prize committee until 1995 to recognize the discovery.

At the time, Dr. Myron Bander of Irvine said the neutrino discovery was "a keystone to our understanding of elementary particle physics."

It was only last June that other scientists, using detectors in Japan, finally determined that the neutrino does have some mass and thus could contribute to the so-called dark matter that pervades and shapes the universe.[12]

From the *Los Angeles Times*:

Martin Perl, who shared the Nobel Prize with Reines for his own work in physics, called his colleague "an energetic man with a fresh view of looking at things."

---

[12] John Noble Wilford, "Frederick Reines Dies at 80; Nobelist Discovered Neutrino," *The New York Times*, August 28, 1998.

> "He always had experiments going in different ways from everyone else," said Perl, a Stanford University physics professor.... "He was always the center of conversations, physics and otherwise," Perl said. He recalled a time the two were at the Moscow airport and became so engrossed in a conversation about physics that they missed their plane.[13]

From *The Independent* (London):

> Reines' physical stature, booming voice and natural, imposing stage presence invariably commanded attention. However, his interactions with people, especially undergraduates, were warmed by his penchant for the lighthearted use of quips, puns, riddles and the poems he was so fond of fashioning and reciting.[14]

Sylvia Reines died in 2006. Afterward Fred and Sylvia's daughter, Alisa Reines Cowden (Lisa), wrote a tribute to her mother. Titled "Creativity and the Life of the Mind," it appeared in a UC Irvine promotional brochure the following year. Her parents, she wrote, shared a deep intellectual curiosity. "My father's study was dense with his achievements, but the rest of the house had my mother's touch. She had impeccable taste and an artist's eye." Lisa noted that paintings and posters hung in every room, which also displayed beautiful sculptures and Indian pottery. Stacks of books were everywhere, and they covered a broad array of subjects, from gardening, literature and history to cooking "of every cuisine." "My mother was familiar with everything in her collection and had read each book at least once."

In speaking with Lisa years later I said that I had thought of Sylvia as clear-headed and direct, a "straight shooter." Lisa responded without hesitation: "Your impression is correct. Straight shooter, that would describe her." She expanded on her view of her parents:

---

[13] Jeff Gottlieb, "Nobel Winner Reines Dies; He Solved a Physics Puzzle," *Los Angeles Times*, August 28, 1998.

[14] William R. Kropp, Jonas Schultz and Henry W. Sobel, "Obituary: Professor Frederick Reines," *Independent* (London), October 19, 1998.

> I really do mean it when I talk about Fred being delightful and playful and looking at the world as full of wonder. Sylvia was sort of the string on his kite, if you want to use that metaphor. She was the grounded one and he was out there with his head expanding into the universe. They were a very good team.[15]

[15] Alisa (Lisa) Reines Cowden interview, 2017.

Chapter 18

The Legacy Grows

During much of Fred Reines' professional life, John Wheeler referred to him as Mr. Neutrino. By the time of Fred's death in 1998, his legacy had been secured not only by his discoveries, but also by the many investigators he had inspired and whose careers he helped advance. Since his death, his legacy has continued to grow. Interest in the ghost particle has soared, as its centrality in the workings of the cosmos has become increasingly evident. Young physicists who never knew Reines have joined the hunt for neutrinos. Nowhere is the expanded interest more evident than in the size and nature of current neutrino-related experiments.

IceCube may be the strangest of all neutrino observatories. Located at the South Pole, its above-ground structures contain computers, data analyzers, and other paraphernalia necessary for its Antarctica investigators. Beneath the surface, the observatory encompasses a huge section of ice whose base lies more than a mile down. Neutrinos and other particles from outer space continuously bombard the Earth. Exactly where these extraterrestrials come from has long puzzled researchers. The observatory was built in quest of answers.

Neutrinos from nuclear reactors, discovered by Fred Reines and Clyde Cowan in 1956, are low energy particles. Their energies are measured in millions of electron volts. In 1962, the Lederman–Schwartz–Steinberger group detected higher energy muon neutrinos generated by the new Brookhaven accelerator. Neutrinos in this accelerator could reach energy levels of 33 billion electron volts. Today, the biggest accelerator on Earth is the Large Hadron Collider (LHC) at CERN, the Geneva-based European Organization for Nuclear Research. The LHC can create particles of 2 trillion electron volts. But IceCube has detected "natural" cosmic neutrinos that dwarf any produced by the LHC. Their energies are measured in quadrillions of electron volts.

In July 2018, Francis Halzen, director of IceCube, announced that an outer-space source of cosmic rays had almost certainly been identified. The source was a type of quasar black hole called a blazar, located at the center of a distant galaxy. Quasars emit powerful jets of radiation that include neutrinos and cosmic rays. When the radiation jets are headed in the direction of Earth, the emitting quasar is described as a blazar. Scientists found that neutrinos passing through IceCube could be traced to a blazar 4.5 billion light years away, in the direction of the constellation Orion.

Construction of IceCube began in 2003 and was completed in 2010. Unlike other underground detectors, which are located in large open spaces, the detection space in IceCube is a section of purely transparent ice. The cube, which measures a half-mile on each side, provides a vast area for detection.

The identification apparatus includes 86 vertical cables. Evenly distanced from each other, they reach down from the Antarctic surface through the cube below. Attached to the cables are more than 5000 photomultipliers (optical sensors). An artist's depiction of a side view suggests a giant diaphanous chandelier.

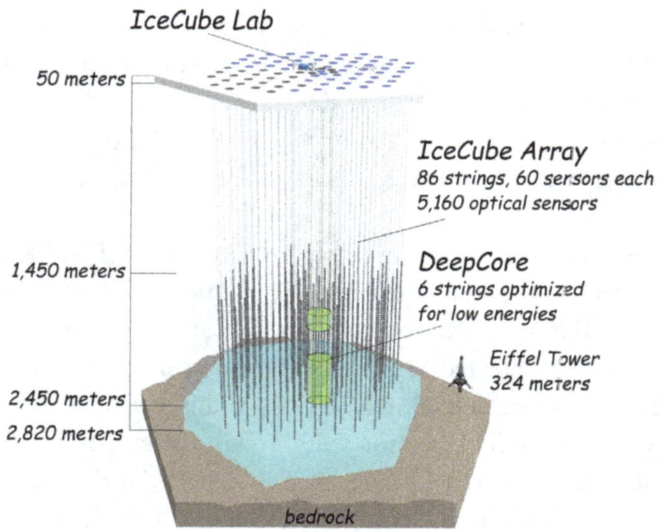

Depiction of IceCube Observatory.

The result is a detector of clear ice and clusters of phototubes that can signal the passage of neutrinos and other particles. Ordinary cosmic rays carry electric charges and their paths may be bent by magnetic fields. But a neutrino has no electric charge and its movement is unaffected by magnetic forces. It travels near the speed of light on a straight path, unimpeded by solid structures. If it interacts with another particle as it passes through the detector, the resulting light emission displays its trail. Thus its straight unwavering path enables the route to be traced back to its source.[1]

Placement of the cables and their photomultipliers required drilling channels through the vast frozen field. Ice melted from the heat of the drilling and refroze around the phototubes. Three hundred physicists from 50 institutions in 12 countries have been part of the IceCube collaboration.[2]

The experiment's director, Francis Halzen, is a professor at the University of Wisconsin, the lead institution for the project. Since its completion the detector, encased in ice and devoid of moving parts, has required little maintenance. Halzen said that operating IceCube can be done from his laptop. In fact, scientists around the world may observe in real time the tracks of particles, on their computer screens. Sometimes only two people remain on site in Antarctica, Halzen said. "Ideally they have nothing to do."[3]

Francis Halzen, https://icecube.wisc.edu/gallery.

[1] Emily Conover, "A High-Energy Neutrino Has Been Traced to Its Galactic Birthplace, *ScienceNews*, July 12, 2018, https://www.sciencenews.org/article/high-energy-neutrinos-blazar-icecube.

[2] "IceCube Explained," https://icecube.wisc.edu/about/overview.

[3] Dennis Overbye, "It Came from a Black Hole, and Landed in Antarctica," *The New York Times*, July 18, 2018.

In the wake of the neutrino-blazar finding, the IceCube collaborators plan to expand the capability of the observatory. They hope to drill new shafts and fill them with more cables and photomultipliers. The addition should lead to identification of more sources of cosmic rays and their associated neutrinos.[4]

The IceCube Observatory has been dedicated to neutrino research since its creation. In contrast, the impetus at other laboratories to engage in neutrino projects grew over time. A prime example is CERN, the nuclear research consortium of European countries. Established in 1954, its creation was prompted by two poignant considerations: the history of European conflict and development of the atomic bomb.

In the first half of the 20th century, rivalries among European powers led to two devastating world wars. The dawning of the atomic age at the end of World War II created the potential for destruction on a scale previously unimaginable. CERN would bring together scientists to collaborate exclusively on non-military nuclear research. The dozen original CERN member states included previous battlefield enemies, notably France and the United Kingdom on one side and Germany and Italy on the other. The chosen research site was a swath of land overlapping the French-Swiss border near Geneva.

Four years after the laboratory was established its first particle accelerator became operational. Since then, the number of accelerators and participating scientists has multiplied. Increasingly powerful accelerators have been built and the number of member states has grown to 22, though about 100 other countries also participate. Some 12,000 scientists now conduct experiments at CERN, which has become the world's biggest particle research facility.

The function of accelerators has essentially remained the same over time: to propel electrically charged subatomic particles at faster and faster velocities, crash them into an object or each other, and assess the resulting breakdown products. As accelerators have become larger, so has their ability to generate higher energy levels.

---

[4] "Neutrino Hunter to Get Upgrade," *Science*, July 26, 2019: 303.

During my visit to CERN in 2018, Michelangelo Mangano, a theoretical physicist there, likened the process to shifting gears in an automobile:

> Getting particles to high energy is a bit like driving a car with a shift stick. You have to shift into first, then second, third, fourth. The same is true with accelerating a particle. You start from rest and accelerate to higher levels of energy.

Michelangelo Mangano, http://mlm.web.cern.ch/mlm/.

In 2008, CERN's Large Hadron Collider began to operate. With its 17-mile circumference it became the world's highest energy accelerator. Protons race around and around gaining energy on each circuit, and are then steered into a collision, resulting in a variety of other products.

A public spotlight shone on CERN in 2012 with reports that the existence of the Higgs Boson had been confirmed. It was the LHC that enabled the finding. The Higgs particle, predicted by the Standard Model, is key to understanding the origin of mass in subatomic particles. Peter Higgs and his colleagues Robert Brout and François Englert proposed the existence of such a particle nearly 50 years earlier. Higgs was in the audience when Fabiola Gianotti, a spokesperson for the ATLAS experiment that was involved with discovery of the Higgs Boson, announced its detection. Amid the cheers that greeted the announcement, Higgs, overcome with emotion, wiped away tears. The following year, for their half-century-old "theoretical discovery," Higgs and Englert received the Nobel Prize. (Brout was no longer alive.)[5]

Earlier neutrino experiments at CERN included NOMAD and CHORUS, which in the 1990s aimed at detecting the tau neutrino as a step toward identifying neutrino oscillations. After Fabiola Gianotti became CERN's director-general in 2016, she spoke of CERN's continuing interest in neutrino research. These "extremely fascinating and elusive particles," she said, could answer "some of the open questions in fundamental physics today."[6]

Current neutrino-related work at CERN includes developing two prototype detectors for DUNE, the international Deep Underground Neutrino Experiment, to be conducted in the United States. In September 2018, the first CERN prototype recorded neutrino tracks. The prototype had taken two years to build and eight weeks to fill with 800 tons of liquid argon. The argon must be maintained at minus 300 degrees Fahrenheit. The prototype is the size of a three-story house. The detector for the actual DUNE experiment, to be hosted by the U.S.'s Fermilab, will be 20 times larger. It is expected to begin recording data in the late 2020s.

Other U.S. national laboratories involved in the DUNE experiment were also originally established to advance peacetime research in atomic energy. The campus of Brookhaven National Laboratory (BNL) includes Michelson Street, Thomson Road, Lawrence Drive, and Rutherford Drive. Named for great physicists of the past, the streets highlight the central role accorded to physics at the Lab. Other streets are named after research universities that were part of the Atomic Energy Agency's consortium that founded the Lab after World War II — Columbia, Johns Hopkins, Harvard, and Pennsylvania.

The sprawling site might be taken for a college campus, but for another unusual feature: horizontal shells in the form of long straight lines

---

[5] "The Brout-Englert-Higgs Mechanism," CERN, https://home.cern/topics/higgs-boson.

[6] Sarah Charley, "Groundbreaking for an International Neutrino Experiment," CERN, July 21, 2017, https://home.cern/news/news/physics/groundbreaking-international-neutrino-experiment.

and large circles. They cover particle accelerators through which subatomic particles are steered by magnets. The straight-line structures cover linear accelerators. Beneath the circular structures are rings through which particles are propelled around a track, gaining energy with each circuit.

BNL was established in 1947 at a former army base on eastern Long Island. Nicholas Samios recalls when each accelerator was built. In 1960, three years after receiving a physics PhD from Columbia University, he began to work at the Brookhaven Lab. He has been there ever since, serving as director from 1982 to 1997. At 86 he remained excited about science.

A blackboard in Samios' office in May 2018 was filled with numbers and Greek letters concerning particle research. Some lettering appeared unrelated to science. Sensing my curiosity, he translated the top line: "May you have many years." He wrote it for a Greek friend, he said, but the expression remained posted, a gesture to all who visited him. Samios smiles easily, as he speaks with pride about the Lab.

By the time Samios arrived on the campus, it had a nuclear research reactor and, by today's standard, a small particle accelerator. Through the years larger machines were built and used in numerous particle experiments. Work at BNL has ranged widely, from developing superconductors and advanced computing methods to the study of space radiation's effects on astronauts.

In 1964, Samios discovered the omega-minus particle, which provided evidence that certain elementary particles were composed of quarks, basic building blocks of matter. He still relishes the long-ago moment.

> We were scanning pictures and I was the guy that saw the particle and measured it. I said to myself that at this instant I am the only one in the world who knows that this thing exists. What a feeling![7]

Samios' enthusiasm is unbridled. "Science is beautiful," he proclaimed. He explained a seemingly impossible feat — how we are able to discern the origin of the universe. His folksy description is straightforward.

> We know now that stuff is made of quarks. A proton is 3 quarks. We want to understand 3 quarks, so how do you do it? In the early 1990s we

---

[7] Nicholas Samios interview, 2018.

> decided to build a machine called the Relativistic Heavy Ion Collider (RHIC). So we'll take 2 huge nuclei, collide them at very high energies, and when they interact, it will be so hot that the protons and neutrons don't exist anymore. You'll have a whole volume of free quarks.
>
> Then as it cools down the quarks will reassemble into particles. So what we will create essentially is the universe at the early time. We had the Big Bang and the quarks and neutrinos, and as it cooled off the quarks made nuclei. Well that's what we've done, in reverse, for the past 18 years.[8]

Samios also spoke of BNL's long association with neutrino research. It was at Brookhaven in 1962 that Lederman, Schwartz, and Steinberger detected the muon neutrino. Raymond Davis joined Brookhaven's chemistry department in the 1950s. His work there on neutrinos was a prelude to his detection of solar neutrinos in the late 1960s.

When the IMB collaboration was created in 1982, Maurice Goldhaber was the sole participant from Brookhaven. The other two collaborating institutions — UC Irvine and the University of Michigan — were each represented by several scientists. Goldhaber had worked in the past with Fred Reines, who brought him into the new collaboration.

Goldhaber was widely respected for his gifted and imaginative contributions to physics. Samios enthusiastically described him as brilliant: "Maurice knew the periodic table of elements in his head." Goldhaber's long association with Brookhaven included 12 years as director beginning in 1961. There he discovered that a neutrino spins in a leftward direction — in the direction a screwdriver turns when it is removing a screw. (Spinning is a characteristic of many subatomic particles.)

Samios was not directly involved with IMB, but he knew Fred Reines from numerous physics meetings they had attended. Fred was very vivacious, he said:

> I remember being at conferences and we'd go out to dinner. At some point Fred invariably would sing. I saw him as a presence. A big guy

---

[8] *Ibid.;* also see Steven Weinberg, *The First Three Minutes: A Modern View of the Origin of the Universe*, 2nd Updated edition. NY: Basic Books, 1993.

with a deep voice. He was exuberant. He was interested in the science and he really liked new ideas.

Nicholas Samios. Courtesy of Brookhaven National Laboratory.

After an hour's conversation, we were joined by Milind Diwan, a senior scientist at Brookhaven. Now 55, Diwan immigrated with his family from India in 1978. He began working at the Lab in 1994, 6 years after receiving a PhD from Brown University. His dissertation had been on the electromagnetic properties of the muon neutrino. He has long been an advocate for the Deep Underground Neutrino Experiment, DUNE, which his website describes as "the highest priority fundamental science project in the U.S."[9]

Diwan was the lead author of a 2003 concept paper, which followed the discovery that neutrinos oscillate and therefore have mass. Still unknown is the actual mass of each of the three forms, or flavors, of neutrinos — electron, muon, and tau neutrinos. Since the Standard Model

[9] "Milind Diwan's Home-Site," https://www.phy.bnl.gov/~diwan/.

only allowed for massless neutrinos, the discrepancy with the fact that neutrinos have mass somehow needed to be addressed. Further, DUNE will also be aimed at resolving a question at the forefront of current physics research: whether matter interacts with neutrinos differently than antimatter.

The paper's two-dozen authors, Samios among them, urged that answers be sought by creating a uniquely large neutrino detector. From the introduction:

> In this paper, we show ... the possibility for an accelerator-based very long baseline experiment that can explore both solar and atmospheric oscillation parameters in a single experiment, complete the measurement of the mixing parameters, and search for a new physics.[10]

The key words were "very long baseline." A baseline is the distance between the source of the neutrinos and the device that detects them; the proposed baseline would need to be greater than for any other accelerator-based neutrino project. The cost of such an experiment was projected at $1.5–2 billion. Funding was not forthcoming for more than a dozen years, and the project was shelved. Finally, with financial backing by the Department of Energy and the National Science Foundation, the 10-year project began in 2017.

With more than 1000 collaborating scientists, excitement about DUNE is keen. Brookhaven's website description is typically impassioned: "The world's most intense neutrino beam will travel hundreds of miles through the Earth, unraveling mysteries and burrowing back in time." Neutrinos are "most often produced in the fiery hearts of stars [and] could be the key to understanding how our universe evolved just after the Big Bang and why the world is made of matter."[11]

---

[10] M.V. Diwan, *et al.*, "Very Long Baseline Neutrino Oscillation Experiments…." *Physical Review*, D 68, 2003, https://arxiv.org/abs/hep-ph/0303081.

[11] "Deep Underground Neutrino Experiment (DUNE)," *Our Science*, Brookhaven National Laboratory, https://www.bnl.gov/science/DUNE.php.

Milind Diwan. Courtesy of Brookhaven National Laboratory.

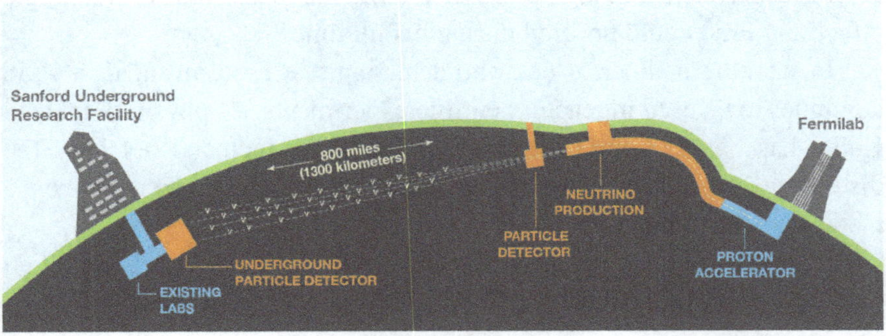

Deep Underground Neutrino Experiment (DUNE), https://www.dunescience.org.

Frustrated by the lengthy delay before funding became available, Diwan and Samios realized that support ultimately must come from members of Congress and White House budget officials, few of whom are scientists. Part of a scientist's mission, they believe, is to educate the public. A perennial question by funders of science projects is: What is the practical value?

Diwan has often been asked the question. His response:

> I would say that there is a difference between a practical value and a cultural value. It is a cultural shift that we know how the sun works deep

> inside. It's in the same class as the Earth going around the sun. It's a shift that occurs in the common man. Oh my God, I now can know what happens inside the sun because 8 minutes later I see it.[12]

Diwan's time projection is based on the fact that solar neutrinos, produced in the sun's core, rise to the sun's surface in seconds. From there they reach Earth in about 8 minutes. Samios nodded agreement and cited Einstein's introduction of relativity as prompting a cultural shift. Diwan then affirmed that "we devalue what we do when we keep insisting on practical uses. Cultural uses should be seen at the same level of importance."

Later in our conversation, Diwan referred to a 2012 article on the apparent disappearance of electron neutrinos. It described a study reconfirming that "the flavor of a neutrino oscillates with time."[13]

The study was, Diwan said, "essentially a repeat of Fred Reines' reactor experiments, though on a much more massive scale. We found the effect that Fred could not find during his lifetime."

In fact, the manner of neutrino detection that Fred invented in 1956 continues to be used in reactor neutrino experiments. As physicist Takaaki Kajita has noted, Reines' seminal contributions included not only the discovery of neutrinos, but also the invention of the delayed coincidence method to detect them.[14]

Examples of such neutrino experiments in recent decades, identified by their acronyms or locations, include:

Daya Bay, China
Chooz and Bugey, (CERN) France
Gosgen, (CERN) Switzerland
NOMAD and CHORUS, (CERN) France-Switzerland
Palo Verde, Arizona
RENO, South Korea

---

[12] Milind Diwan interview, 2018.

[13] F.P. An, *et al.*, "Observation of Electron-Antineutrino Disappearance at Daya Bay." *Physical Review Letters*, 108, April 2012, https://journals.aps.org/prl/abstract/10.1103/PhysRevLett.108.171803.

[14] Takaaki Kajita, email communication, 2019.

Each of these experiments included numerous collaborators from various institutions.

The DUNE project further echoes the fundamentals of Fred's neutrino research efforts: first, identify or generate a source of neutrinos; then, build an apparatus of sufficient size, design, and content to detect a neutrino interaction.

By 2019, DUNE was already the world's largest neutrino-experiment collaboration. Participating institutions include CERN, Brookhaven, and other laboratories around the world. None is more central to the project than Fermilab in Illinois, an hour's drive from Chicago. Many scientists there, as at other participating facilities, are also based at different universities and research institutions. As noted previously, Fermilab's website states that it is the neutrino capital of the world. Boris Kayser, a veteran physicist there, agrees: "Much to my happy surprise, neutrino experimental research has become *the* major thrust of Fermilab."[15]

The facility was established in 1967, 13 years after Enrico Fermi's death. It is a national laboratory whose 1750 employees study particle physics. The Lab's vision is "to solve the mysteries of matter, energy, space and time for the benefit of all." Management is by a consortium of nearly 100 research universities, headed by the University of Chicago. Using the Lab's particle accelerators and colliders, scientists and engineers have engaged in a range of experiments on the building blocks and forces that govern the universe.[16]

Stephen Brice received his PhD in physics from Oxford University in 1996. Five years later he joined Fermilab and now heads its Neutrino Division. In September 2018, Brice graciously hosted me at the Lab and at meetings with managers of various neutrino projects. At one point we entered a nondescript building and were greeted by Louise Suter, the young manager of operations for several detectors. She glowed with enthusiasm about her work. Her 2012 PhD from England's University of

[15] Kayser interview, 2018; http://www.fnal.gov/pub/science/particle-physics/experiments/neutrinos.html.

[16] "Fermilab Is America's Particle Physics and Accelerator Laboratory," http://www.fnal.gov/pub/about/index.html.

Manchester was based partially on research that she had conducted at Fermilab.

Stephen Brice. Neutrino Division Head, Fermilab.

Louise Suter. Manager, Neutrino Detectors, Fermilab.

Inside the building two large screens displayed lines, circles, arcs, and some odd squiggles. The images changed every few seconds, replaced by new configurations at different angles. We were seeing tracks caused by particles and rays passing through two detectors. One detector was 400 miles distant in northern Minnesota and the other detector, which we were about to visit, was 350 feet beneath us. Both are part of NOvA, an acronym for the two-detector neutrino experiment (NuMI Off-Axis ν_e Appearance).

The screen images are also visible in real time in laboratories around the world. Neutrinos emitted from a Fermilab accelerator first pass through the near-detector and milliseconds later enter the far-detector in Minnesota. The aim is to observe oscillations during a particle's passage and to search for a possible fourth "sterile" neutrino. NOvA and several other experiments have been forerunners of DUNE, which will dwarf every other accelerator neutrino experiment to date.

Farther inside the building is an elevator. It descends to the underground cavern and tunnels that were hollowed out to house the near-detector and other research apparatus. Before boarding, visitors must view a video about safety. The narrator begins with a warning: When entering the underground areas "you need to be aware of the potential hazards and understand the procedures we have put in place to mitigate them."

Spurts of flying rock, which the narrator refers to as "dropping pebbles," require hard-hat protection.

> You will get a hard hat just before getting on the elevator. Also, the MINOS detector is an electromagnet that may be energized during your tour.... If you have a pacemaker you must stay out of the magnetic field area. [MINOS is an acronym for Main Injector Neutrino Oscillation Search.]

The narrator assures viewers that oxygen masks are available if the atmosphere becomes hazardous. Despite the warnings, actual conditions did not seem at all threatening. The several minutes-long elevator ride, though slow and bumpy, was entirely casual. Upon reaching the bottom, Steve Brice, Louise Suter, and I entered a cavernous rock-lined hall. The temperature was comfortably cool. I thought of the starkly contrasting conditions for neutrino hunters here and those in the perpetually uncomfortable 1960s-South African mine laboratory.

As we walked along rows of plumbing and wiring, tubes and cables, Suter and Brice mentioned that numerous experiments had been conducted there. Among them was MINOS, a 9-year project that ended in 2014, although its detector is still sometimes energized for current use. It had been the first Fermilab-to-Minnesota neutrino experiment and was the basis for scores of published studies. The number of authors of any one study commonly exceeded 100. Sample titles included:

"Constraints on large extra dimensions from the MINOS experiment"

"Search for sterile neutrinos mixing with muon neutrinos in MINOS"

"Observation of seasonal variation of atmospheric multiple-muon events in the MINOS near and far detectors"

"New constraints on muon-neutrino to electron-neutrino transitions in MINOS"[17]

Gleaning more information about neutrino oscillations would require a far larger research design than for previous accelerator neutrino experiments. Thus, the current focus on developing DUNE.

The new DUNE project includes advancing the capability of producing high-energy neutrinos so that a huge flux may be beamed from Fermilab. The swarm of neutrinos will be aimed toward an underground detector 800 miles away. After passing into and out of the near detector the particles will travel through the Earth like bullets through fog. The distant detector will be located in the Homestake mine in South Dakota where Raymond Davis detected solar neutrinos 50 years earlier. The concentration of neutrinos and the long baseline will greatly increase the likelihood of seeing neutrino oscillations as well as other interactions.

The new detector in South Dakota, one mile beneath the earth's surface, will be mammoth. To create the required space, drilling and dynamite must tear away 875,000 tons of rock, equivalent to the mass of eight large aircraft carriers. Loosening and lifting the rock from the mine and then depositing it in a mile-long quarry will take about three years. The excavation should be completed in the early 2020s. Completion of the giant DUNE detector in the newly provided space is expected in 2026, or later.[18]

In the course of our meetings at Fermilab, Brice and I spoke with Boris Kayser. A theoretical physicist, Kayser, then 80, joined the Lab in 2001. He is well-known in the physics world. Before Fermilab he spent 29 years at the National Science Foundation in charge of funding theoretical physics projects. He also continued his own research, "which as a theorist

---

[17] Minos Experiment Publications, http://www-numi.fnal.gov/Minos/minospub.txt. Also see Don Lincoln, "How Do You Make a Neutrino Beam?" https://www.youtube.com/watch?v=y8kJFXhN7Po.

[18] "Dune at Long-Baseline Neutrino Facility," Fermilab, https://lbnf.fnal.gov/cf.html

I could do because I don't need a laboratory." Kayser is intrigued by issues at the frontier of particle physics including the relationship between matter and antimatter, but he finds no project more exciting than DUNE. It promises to further divulge the nature of neutrinos and advance an understanding of why we are here.

Boris Kayser. Theoretical Physics Department, Fermilab, https://theory.fnal.gov/people/emeriti/.

At conferences over the years, Kayser interacted with Fred Reines and found him charming.

> He had a personality that was very distinctive. He had what I would describe as a low-key intensity. If you had a conversation with him it was outwardly low key. But you could sense that behind it he was very focused, especially if it was about physics.[19]

In view of Reines' fondness for big ideas, I wondered how he might feel about the number of huge neutrino projects now underway. Kayser guessed that Fred would react as many in the physics community do: "We would not have imagined a decade or two ago research into neutrinos would be as big a deal today as it has become." Perhaps Kayser is correct.

[19] Boris Kayser interview, 2018.

Still, a hallmark of Fred's reputation was his inclination toward large and imaginative projects. Several never materialized, but many did. For much of his career his intensity and persistence resulted in productive outcomes. As a result, some colleagues believed that Fred had a sixth sense.

The arc of Fred's successes springs from his expressed ambition as a teenager to do something great. Its fulfillment continues to be realized long after his passing.

Epilogue

Prior to 2015, the only memorial feature in the spacious lobby of Frederick Reines Hall was the bust of Reines. But that year a collection of memorabilia was added as a permanent exhibit. It contains more than 100 items including Fred's Eagle Scout badge, a hard hat worn in deep-mine laboratories, a copy of his Nobel Prize announcement, photographs, scientific articles, essays, poetry, and various correspondence. They are displayed in glass showcases, artfully arranged throughout the area.

Upon entering the lobby, a visitor encounters a compelling visual, on the right: a 40-foot tall photograph of Japan's Super-Kamiokande (Super-K) detector. Burrowed deep under Mount Ikeno, the cavernous structure inspires wonder. The detector's inside cylindrical wall is 138 feet tall, about the height of a 13-story building.

Covering the Super-K's inner surfaces are some 11,000 photo tubes. From every angle a viewer sees the tubes as clusters of uniformly positioned "stars." The floor tubes are under a low level of water that supports a small floating platform on which three workers in safety outerwear are standing, dwarfed by the surrounding vastness. When the entire cavern is filled with 50,000 tons of ultra-pure water, traversing cosmic particles can generate pulses of light that may signal a neutrino interaction.

A description adjacent to the Super-K photograph indicates that the detector is operated by an international collaboration of scientists. The description also notes that principal Super-K figures are scientists from the UC Irvine "Neutrino Group founded by Fred Reines."

Henry Sobel and other members of the Neutrino Group are also part of a mammoth successor project to Super-K. At a cost of $660 million, nearby Hyper-Kamiokande will be 10 times larger. When completed in the late 2020s, "it will be able to capture that much more data about neutrinos emanating from the Sun, distant stars, and supernovae."[1]

Kim Weiss received a PhD in the History of Science from UC Irvine in 2013. She later began to work with William Kropp to create the Reines Hall exhibit. After combing through the library's Reines Collection, and speaking with other colleagues of Reines, they completed the project in 2015. Kim never met Fred, but over time she felt that she knew him. "He was larger than life," she said. "Once he had an idea, he seems to have carried it through regardless of any opposition, whether scientific, financial, political or anything else." She was pleased that students who see the exhibit will learn about "this endearing person who, since his early years, examined what kind of person he wanted to become."

The Super-Kamiokande Image on the Lobby Wall of Reines Hall.

[1] "Japan Boosts Neutrino Efforts," *Science*, 367 (6473) January 3, 2020: 8.

Part of the display in the Reines Hall Lobby, Courtesy of Richard Lichtenberg.

The new exhibit was not the only 2015 celebration of Fred's life and accomplishments. An annual lecture series was also inaugurated that year. As described on the university's website:

> The Reines Lecture Series brings world renowned physicists and astronomers to UC Irvine to give a public lecture and a department colloquium. This celebration honors the rich history of the Department of Physics & Astronomy and brings together faculty, students, supporters, and the community at large.[2]

Apropos of the description, the initial lecturer in 2015 was Frank Wilczek. He was succeeded by Steven Chu in 2016, Arthur McDonald in 2017, Kip Thorne in 2018, and Jocelyn Bell Burnell in 2019. The first four

[2] Reines Lecture Series, https://ps.uci.edu/rls.

lecturers were Nobel laureates whose work included contributions to the discovery of the strong force (Wilczek), methods to cool and trap atoms (Chu), neutrino oscillations (McDonald), and gravitational waves (Thorne). As a graduate student in 1967, Bell Burnell discovered pulsars — highly dense neutron stars that emit pulses of radiation. A Nobel Prize was awarded for the discovery to Bell Burnell's adviser, Antony Hewish, but not to her. As is common practice by the Nobel Committee, student researchers are not given the prize despite their contribution to a discovery. Still, Bell Burnell has received many other awards both for her pulsar discovery and for decades of leadership in the field of astrophysics.

While the expertise of the Reines lecturers ranged over diverse fields, several noted in their remarks the high esteem accorded to Fred in the overall physics world. The exceptional caliber of the speakers has positioned the Reines Lecture Series as a notable annual scientific event.

The following year, another event added further luster to the Reines legacy. The Los Alamos Historical Society opened a gallery in the cottage where Hans Bethe's family lived during the Manhattan Project. Prepared in cooperation with the Atomic Heritage Foundation, its exhibits are largely about Los Alamos during the Cold War. A wall of the Hans Bethe House lists the names of the 24 Nobel laureates who have been associated with the Los Alamos National Laboratory. Fred Reines is the only laureate whose award was for work he developed while based at the Lab. In acknowledgement of this distinction, a copy of his Nobel medal and a plaque citing his neutrino discovery are exhibited in a large glass case.

One additional item regarding Fred is exhibited in a nearby showcase. It is a program of the 1955 local production of *Finian's Rainbow*, in which Fred played a leading role. When I visited the Hans Bethe House in 2018, I wondered why among the hundreds of available Reines artifacts, only the show's program was displayed. Heather McClenahan, executive director of the historical society said, "We just thought it was a nice example of Fred's diverse talents."

Heather did not know much about the play. I explained that I was familiar with the storyline and found interesting parallels with Fred's real-life experience. She listened intently as I said why her choice seemed providential.

The play was staged in Los Alamos during the year prior to the Reines–Cowan detection of reactor-generated neutrinos. For years afterward, the play seemed to remain in Fred's thoughts. The theme included a search for gold underground at the end of a rainbow. Fred later sought his "gold" — natural neutrinos — beneath the ground in a gold mine. The title of his course for non-science majors, "Rainbows and Things," was itself a reminder of the play. I told Heather that besides describing refraction and the science behind rainbows, he told his students that rainbows could be a source of hope and inspiration. I repeated Fred's words: "There's a whole legend about this," he had told his class. "You see a rainbow in the sky and you say: Things are looking better now."[3]

I also told Heather that she might find surprising the lyrics to the show's classic song, *Look to the Rainbow*. Toward the end of his 1955 performance Fred's character, Woody, sang the song in a duet with his romantic partner, Sharon. Excerpts from the text seem an allegory of Fred's lifelong dream: to be an extraordinary physicist and to gain deserved recognition.

Look to the Rainbow

Look, look, look to the rainbow.
Follow it over the hill and stream.
Look, look, look to the rainbow.
Follow the fellow who follows a dream....

So, I bundled me heart and I roamed the world free.
To the east with the lark.
To the west with the sea.
And I searched all the world and I scanned all the skies.
But I found it at last in my own true love's eyes.
[Songwriters: Burton Lane and E.Y. Harburg]

During Fred's life he "roamed the world free" and he "scanned all the skies." He kept searching metaphorically for a rainbow, which could point to the treasure he so persistently sought. He actually *was* "the fellow who follows a dream." The "true love" that his character in the play found was his romantic partner, Sharon. Outside the play it was also fulfillment of his enduring search for knowledge about the neutrino.

The Nobel Prize in 1995 would seem a crowning conclusion to his youthful ambition, "to discover one little thing and build my career around it." But until Fred's passing 3 years later, he remained devoted to learning still more about the elusive particle.

Interest in neutrino studies continues to grow. A biennial conference on neutrinos was scheduled to take place in Chicago, in June 2020.

[3] Transcription, in Reines Collection.

Because of the COVID-19 pandemic, it was held as a virtual event via the internet, with talks, posters, and interactive discussion. Its 3400 attendees from around the world quadrupled the number ever at a previous neutrino conference. Fermilab's Steve Brice, a conference organizer, attributed the increase in part to heightened excitement about the subject. Investigators, he observed, enjoy the challenge of uncovering the intriguing secrets held by the elusive particles. There is physics beyond our current understanding of the universe, he said. "Pursuing neutrinos is one of our best hopes of learning more of these fundamental truths about our world."[4]

Fred Reines would surely have been pleased by the expanding interest in neutrinos, and especially by the resolute encouragement of their continued pursuit.

Courtesy of the AIP Emilio Segrè Visual Archives.

[4]"Neutrino 2020 Kicks Off June 22," Fermilab, https://www.eurekalert.org/features/doe/2020-06/dnal-n2k061720.php; "More Than 3,000 Neutrino Scientists Gather Online for Neutrino 2020," Fermilab, https://news.fnal.gov/2020/07/more-than-3000-neutrino-scientists-gather-online-for-neutrino-2020/.

Appendix

An Endearing Reminiscence

Robert Hofstadter won the 1961 Nobel Prize in physics. In 2020 Ken Ford introduced me to Robert's son, Douglas Hofstadter, author of the Pulitzer Prize-winning science classic, *Gödel, Escher, Bach*. Douglas then shared with me an endearing reminiscence, not previously published, that he had composed weeks after connecting with Fred Reines at the 1995 Nobel Prize ceremony.

Recollections of Fred Reines by Douglas Hofstadter

Amidst a flurry of social activities involving the 1995 Nobel Prize-winners, the press, and international guests like myself, I heard rumors that Fred Reines had actually made it to Stockholm after all. That thought pleased me. I had a warm memory of him from many years earlier, and it had struck me as very sad if he were not able to attend this marvelous occasion honoring his greatest achievement, which to me had a very special personal meaning, because as a young boy, I had been totally fascinated with neutrinos (at that time there was only one species of neutrino that was known, along with its antiparticle; today, there are three!), and when I was in third grade, I had even fantasized about *being* one! I'd picked up little snippets about various particles from my Dad, and in my kidlike way, I'd invented some variant of tag in which my little friends zipped about the playground being electrons and positrons and photons, while I got to savor the self-assigned glory of being a neutrino!

It was fun for us all, though of course we had barely any inkling of what these mysterious micro-entities were. Many years later, as a graduate student in physics, I was still fascinated by neutrinos more than by any other particle. And so, for this and for other reasons, I was very glad to hear that Fred Reines might actually be in town.

That afternoon, in a palatial room in some vast, majestic Stockholm edifice, there was a swanky reception for several hundred people decked out to the nines, savoring (none too subtly) seeing and being seen, suavely sipping from svelte, tapered flutes filled with sparkling champagne, and carefully sifting from the huge *smörgåsbord* a sampler of succulent Scandinavian snacks. And then, all of a sudden, to my total surprise, I spotted Fred Reines in the crowd. He *had* in fact made it across the ocean! How terrific! I felt very happy for him. Moreover, at that moment, against all odds, no one was talking with him. Quickly taking advantage of this unexpected lull, I walked up and introduced myself, reminding him that we'd once met, long ago and far away in Washington, D.C., and of course congratulating him on his honor.

From the get-go, Professor Reines was exceedingly cordial, even sweet. Not only did he say he remembered me and was glad to see me again, he also asked with evident interest whatever had brought me to the Nobel festivities. I explained my curious status as a "Distinguished International Guest", which intrigued him, and then I told him how pleased I was that he'd made it to Stockholm, since I'd heard his health might have prevented him from doing so. He smiled and said it hadn't been too easy for him, but that luckily he'd managed to arrive safe and sound a couple of days earlier, and was very grateful to be present and to be participating in these festive events. And then, quite out of the blue, he told me he'd always been very fond of my Dad, and for a long time he'd wondered why my Dad had moved from Princeton to Stanford and what had impelled him to take up electron scattering when he got there.

As it happened, I knew the answers to these questions very well, so I plunged into the story of how Princeton's Physics Department had been painfully slow in promoting my Dad, and how Felix Bloch and Leonard Schiff, at Stanford, sensing some potential in this injustice, had "wooed" my Dad from his native East Coast all the way out to the Wild West, and how, on our family's cross-country move in the summer of 1950, we had stopped for one night in Saint Louis, where my Dad's old friend Eugene Feenberg, a physicist at Washington University, had strongly encouraged him to take advantage of Stanford's newly-built linear accelerator in order to probe the inner structure of atomic nuclei, and how this chance conversation in Saint Louis had had a major effect on my Dad's research in those key early days of his career.

Reines was clearly fascinated, and then, not yet done with me, he said, "And tell me, how is Bob Herman doing these days?" Now this was another topic that I knew inside-out, for Bob was one of my Dad's oldest and dearest friends (they'd met at age 16 in New York), and I'd known him ever since I was a tiny kid. Bob was also a very creative and pioneering physicist, as well as one of the funniest people I'd ever met, and he had recently written a loving memoir of my Dad to serve as a preface to my Dad's collected papers, to which I had added some of my own remarks, and then helped him polish his preface's style, which Bob had greatly appreciated. So I related a couple of colorful anecdotes about Bob, which tickled Fred Reines pink. What fun!

At this point, we'd been conversing for ten minutes or so and I was starting to feel a bit guilty for taking up too much of Professor Reines's precious time at this momentous event, so I said I didn't want to keep him from circulating, but he graciously replied that he was in no hurry and would love to hear more about Bob Hofstadter and Bob Herman. Of course I was flattered, so I obliged him with another brief story or two, but then I again started feeling twinges of guilt, this time because I'd been talking much more than he had, and so, trying to draw him out a bit, I casually tossed a rather silly question his way: "Tell me — of all the things you've done so far here in Stockholm, what have you enjoyed the most?"

I had no idea what to expect, but I was absolutely and totally floored when Fred Reines said to me, "Oh — talking with you!" I couldn't believe my ears. Talking with me?! How could this possibly be the case? Here was Frederick Reines, at the acme of human glory, being exalted to the skies by Swedish royalty, surrounded by some of the most brilliantly shining stars in the vast intellectual firmament, and he was telling me that this simple conversation with "little ol' me" was the most enjoyable thing of all so far? Wow! What a mensch he was!

This wildly improbable offhand remark — spontaneous, generous, glowing with human warmth — imprinted itself indelibly in my memory, and in an eyeblink made my entire trip worth it. Those three simple words uttered by Fred Reines, co-discoverer of the first neutrino and exemplar of openness and kindness, remain a highlight of my life.

Acknowledgements

Much of this book is a product of the wisdom, experience, and memories of many individuals. None were more understanding of Fred Reines and his devotion to science, especially to neutrino studies, than those who worked with him. Some of the more than 50 physicists with whom I spoke, or whose writings or videos provided relevant information, were especially close to Fred. Several worked with him for decades. William Kropp, who viewed Fred as his scientific father, spent countless hours with me by phone, email, and in person reviewing his nearly 40-year association with Fred. Henry Sobel's physics career, like Bill Kropp's, began in the 1960s as a graduate student of Fred's at Case Institute and continued with him at Irvine. Hank was also unfailingly generous with his answers to my many questions.

Harris Mayer recounted his lengthy and close association with Fred beginning soon after World War II, when both worked at the Los Alamos National Laboratory. To Kenneth Ford I am especially grateful, for sharing the many facets of his history with Fred. They met during the 1950s in Los Alamos and grew closer after 1966 at the newly established University of California, Irvine. Ken Ford was the founding Chair of the school's Department of Physics and Fred the founding Dean of Physical Sciences. Also, Ken read my entire manuscript and his comments helped immensely to ensure scientific accuracy and clarity.

William Kropp, 2018.

Henry Sobel, 2018.

Harris Mayer, 2018, Courtesy of the Atomic Heritage Foundation.

Kenneth Ford, 2018.

Other physicists whom I interviewed were also part of Fred's Neutrino Group at UC Irvine including Peter Doe, Henry Gurr, Todd Haines, Michael Moe, and LeRoy Price. Jonas Schultz led his own research team, though he collaborated with the Neutrino Group in the 1980s, on the IMB experiment. The "Frederick Reines Papers" at the UC Irvine library (referred to in the text as part of the "Reines Collection") is an invaluable source of information about Fred. Besides correspondence and other papers, it contains artifacts and audio-visual materials. I thank Audra

Eagle Yun, Derek Quezada, and Krystal Tribbett for facilitating my access to the collection. Science historian Kim Weiss provided valuable guidance in sorting through the material.

For material about Fred during his student years at the Stevens Institute of Technology, I thank Ted Houghtaling, archivist at the school's library. Additional information about Fred is available at the Los Alamos Historical Society and the Los Alamos National Laboratory library. For help in accessing materials at these facilities I thank Alan Carr, Don Cavness, Rebecca Collinsworth, Matthew Hopkins, Heather McClenahan, and John Moore. Scientists currently or formerly at the Los Alamos Lab, who helped with my inquiries about Fred and neutrinos, include physicists Richard Garwin, Steven Greene, Todd Haines, William Louis, Harris Mayer, and cartographer Andrea Kron. Cynthia Kelly, founder of the Atomic Heritage Foundation, and associate of the National Museum of Nuclear Science and History, offered valuable perspectives on the Atomic Age and its legacy.

At Brookhaven National Laboratory, I met with physicists Nicholas Samios and Milind Diwan whose insights and recounting of experiences with Fred were particularly helpful. I also heard summaries by others there whose work has included neutrino-related research: Thomas Tsang, Sergio Rescia, and Robert Palmer.

I visited the CERN particle physics laboratory in Geneva as part of the 2018 *New York Times Journeys* program. Participants received briefings on the Large Hadron Collider and various CERN experiments, some involving neutrino research. Physicist-presenters with whom I spoke included Michelangelo Mangano, James Gillies, Marta Bajko, and Emma Sanders.

At Fermilab, Boris Kayser offered insightful perspectives about Fred, with whom he had numerous interactions. Stephen Brice, head of the Neutrino Division at the Lab, graciously hosted me there and arranged meetings with other scientists and Lab personnel including Valerie Higgins, Alison Markovitz, Louise Suter, and Peter Wilson.

Scientists from other institutions who had worked with Fred or were otherwise familiar with his work, and with whom I communicated, include Neta Bahcall, Christine Chen, Gary Feldman, Takaaki Kajita, John LoSecco, Arthur McDonald, and Lawrence Sulak. Historians

Beverly Flanagan and Marianne Leese provided archival records about Hillburn, New York. Others who helped with information relevant to the manuscript include Catherine Chen, George Daly, Joanne Ford, Douglas Hofstadter, Christa Kropp, Avi Naiman, Howard Peskoe, Susan Peskoe, Jennifer Prost, Marcia Rudy, and David Schwartz.

I thank relatives and colleagues of Clyde Cowan for their recollections about Clyde and their impressions of the Reines-Cowan team's discovery of the neutrino. They include Clyde's son, George Cowan, his daughter, Elizabeth Cowan Riordon, and grandchildren David Riordon and James Riordon. Physicist Hall Crannell recounted his relationship with Clyde when both were professors at Catholic University. I also thank librarian Shane MacDonald for enabling my access to the Cowan archives at the Catholic University library.

Finally, I am grateful for the observations of several of Fred's and my family members. They include his daughter Alisa Reines Cowden and cousins and nieces whose frequency of interactions with Fred varied. Even those whose personal contacts with him were minimal if any, shared reflections derived from their parents' or other family members' experiences. They include Norman Corwin, Carolyn Reines-Graubard, Barry Graubard, Tamara Kaiser, Anita Lichtenberg, Richard Lichtenberg, Arthur Sapper, Joan Rogers, Philip Cole, Wendy Cole, William Cole, and Ruth Cole. My own conversations with Fred through the years, and about him with now-deceased relatives, also enhanced my understanding of his persona. This was especially the case with Fred's brothers William and David and their mother, my aunt Gussie Reines.

Endnotes

The term "Reines Collection" here refers to a special collection of items in the UC Irvine Library, and images provided by William Kropp and Kim Weiss. The library collection, titled "Reines (Frederick) Papers circa 1929–2007," is described as documenting "Frederick Reines' career in nuclear physics and astrophysics as both a scientist and an academic, as well as aspects of his personal life."

"Nobel Lecture" refers to Reines' presentation at the 1995 Nobel Prize ceremony.

Interviews were conducted variously in person, by telephone, and via email.

Index

CPSIA information can be obtained
at www.ICGtesting.com
Printed in the USA
JSHW021437080721
16675JS00001B/63

9 789811 231056